THE INTERNATIONAL
ENCYCLOPEDIA
OF PHYSICAL CHEMISTRY
AND CHEMICAL PHYSICS

Topic 2. CLASSICAL AND QUANTUM MECHANICS

EDITOR: PROFESSOR R. MCWEENY

Volume 4

QUANTUM THEORY OF SCATTERING PROCESSES

BY

J. E. G. FARINA

THE INTERNATIONAL ENCYCLOPEDIA OF PHYSICAL CHEMISTRY AND CHEMICAL PHYSICS

Members of the Honorary Editorial Advisory Board

J. N. Agar, *Cambridge*
R. M. Barrer, *London*
C. E. H. Bawn, *Liverpool*
N. S. Bayliss, *Western Australia*
R. P. Bell, *Stirling*
C. J. F. Böttcher, *Leyden*
F. P. Bowden, *Cambridge*
G. M. Burnett, *Aberdeen*
J. A. V. Butler, *London*
C. A. Coulson, *Oxford*
J. S. Courtney-Pratt, *New Jersey*
D. P. Craig, *Canberra*
F. S. Dainton, *Oxford*
C. W. Davies, *London*
B. V. Derjaguin, *Moscow*
M. J. S. Dewar, *Texas (Austin)*
G. Duyckaerts, *Liège*
D. D. Eley, *Nottingham*
H. Eyring, *Utah*
P. J. Flory, *Stanford*
R. M. Fuoss, *Yale*
P. A. Giguère, *Laval*
W. Groth, *Bonn*
J. Guéron, *Brussels*
C. Kemball, *Edinburgh*
J. A. A. Ketelaar, *Amsterdam*
G. B. Kistiakowsky, *Harvard*
H. C. Longuet-Higgins, *Edinburgh*
R. C. Lord, *Massachusetts Institute of Technology*
M. Magat, *Paris*
R. Mecke, *Freiburg*
Sir Harry Melville, *London*
S. Mizushima, *Tokyo*
R. S. Mulliken, *Chicago*
R. G. W. Norrish, *Cambridge*
R. S. Nyholm, *London*
J. T. G. Overbeek, *Utrecht*
K. S. Pitzer, *Stanford*
J. R. Platt, *Chicago*
G. Porter, *The Royal Institution, London*
I. Prigogine, *Brussels (Free University)*
R. E. Richards, *Oxford*
Sir Eric Rideal, *London*
J. Monteath Robertson, *Glasgow*
E. G. Rochow, *Harvard*
G. Scatchard, *Massachusetts Institute of Technology*
Glenn T. Seaborg, *California (Berkeley)*
N. Sheppard, *East Anglia*
R. Smoluchowski, *Princeton*
H. Stammreich, *São Paulo*
Sir Hugh Taylor, *Princeton*
H. G. Thode, *McMaster*
H. W. Thompson, *Oxford*
D. Turnbull, *G.E., Schenectady*
H. C. Urey, *California (La Jolla)*
E. J. W. Verwey, *Philips, Eindhoven*
B. Vodar, *Laboratoire de Bellvue, France*
M. Kent Wilson, *Tufts*
Lord Wynne-Jones, *Newcastle-upon-Tyne*

THE INTERNATIONAL ENCYCLOPEDIA OF PHYSICAL CHEMISTRY AND CHEMICAL PHYSICS

Editors-in-Chief

D. D. ELEY, NOTTINGHAM

F. C. TOMPKINS, LONDON

List of Topics and Editors

1.	Mathematical Techniques	H. Jones, *London*
2.	Classical and Quantum Mechanics	R. McWeeny, *Sheffield*
3.	Electronic Structure of Atoms	C. A. Hutchison, Jr., *Chicago*
4.	Electronic Structure of Molecules	J. W. Linnett, *Cambridge*
5.	Molecular Structure and Spectra	Editor to be appointed
6.	Kinetic Theory of Gases	E. A. Guggenheim, *Reading* (Deceased)
7.	Classical Thermodynamics	D. H. Everett, *Bristol*
8.	Statistical Mechanics	J. E. Mayer, *La Jolla*
9.	Transport Phenomena	J. C. McCoubrey, *Birmingham*
10.	The Fluid State	J. S. Rowlinson, *London*
11.	The Ideal Crystalline State	M. Blackman, *London*
12.	Imperfections in Solids	J. M. Thomas, *Aberystwyth*
13.	Mixtures, Solutions, Chemical and Phase Equilibria	M. L. McGlashan, *Exeter*
14.	Properties of Interfaces	D. H. Everett, *Bristol*
15.	Equilibrium Properties of Electrolyte Solutions	R. A. Robinson, *Washington, D.C.*
16.	Transport Properties of Electrolytes	R. H. Stokes, *Armidale*
17.	Macromolecules	C. E. H. Bawn, *Liverpool*
18.	Dielectric and Magnetic Properties	J. W. Stout, *Chicago*
19.	Gas Kinetics	A. F. Trotman-Dickenson, *Cardiff*
20.	Solution Kinetics	R. M. Noyes, *Eugene*
21.	Solid and Surface Kinetics	F. C. Tompkins, *London*
22.	Radiation Chemistry	R. S. Livingston, *Minneapolis*

QUANTUM THEORY OF SCATTERING PROCESSES

BY

J. E. G. FARINA

LECTURER IN APPLIED MATHEMATICS IN THE UNIVERSITY OF NOTTINGHAM, ENGLAND

PERGAMON PRESS

OXFORD · NEW YORK · TORONTO

SYDNEY · BRAUNSCHWEIG

Pergamon Press Ltd., Headington Hill Hall, Oxford
Pergamon Press Inc., Maxwell House, Fairview Park, Elmsford, New York 10523
Pergamon of Canada Ltd., 207 Queen's Quay West, Toronto 1
Pergamon Press (Aust.) Pty. Ltd., 19a Boundary Street, Rushcutters Bay, N.S.W. 2011, Australia
Vieweg & Sohn GmbH, Burgplatz 1, Braunschweig

First Edition 1973

Library of Congress Cataloging in Publication Data

Farina, John Edward George, 1933—
Quantum theory of scattering processes.

(The International encyclopedia of physical chemistry and chemical physics. Topic 2: Classical and quantum mechanics, v. 4)

1. Scattering (Physics) 2. Collisions (Nuclear physics) 3. Quantum theory. I. Title. II. Series: The International encyclopedia of physical chemistry and chemical physics (Oxford) Topic 2: Classical and quantum mechanics, v. 4.

QD453.15 topic 2. Vol. 4 [QC794] 539.7′54 72–10162
ISBN 0–08–017047–1.

PRINTED IN GREAT BRITAIN BY BELL AND BAIN LTD., GLASGOW

CONTENTS

PREFACE

HISTORICALLY the subject of quantum scattering theory developed soon after the discovery of the basic principles of quantum mechanics with the formulation of the method of partial waves by Faxen and Holtsmark and the approximation of Born. The first book on the subject (the famous monograph of Mott and Massey) appeared soon after this. The rapid development of nuclear and high energy physics stimulated a transformation of the subject as a result of the work of Schwinger, Lippmann and others, and these changes were incorporated and further developed in the book of Goldberger and Watson.

For many years the only book on the subject was that of Mott and Massey. The increasing importance of scattering processes in atomic and molecular physics as well as in nuclear and high energy physics has led recently to a rapid increase in the number of texts on quantum scattering. These tend to be either of an advanced and formal nature which makes them difficult for beginners, or they are of a fairly specialized nature. An exception to this general rule is the book of Rodberg and Thaler, which gives an elementary account of the basic ideas of the subject but does not contain many examples from atomic and molecular physics.

The present volume aims to be an introduction for physicists, chemists, mathematicians and others to the quantum theory of collisions. It concentrates on basic concepts and principles rather than actual techniques of calculation, although the latter are of course discussed, partly for their own sake and partly to illustrate the theory. Examples are taken from molecular and atomic physics. The knowledge presupposed on the part of the reader is an understanding of the basic principles of quantum mechanics as contained in, for example, Volume 1.‡

The first two chapters deal with the basic ideas of time-independent scattering theory as applied to the simple situation of the scattering of a single particle by a centre of force, while Chapters 3 and 4 extend these ideas to the more complicated cases of the scattering of two simple particles, particles with structure, and rearrangement collisions.

‡ References to companion volumes in Topic 2 of the *International Encyclopedia of Physical Chemistry and Chemical Physics* are indicated simply by giving the volume number.

Chapter 5 takes up the exposition of formal scattering theory, whose growth represents one of the important developments in physics since the war. In Chapter 6 the time-dependent treatment of scattering of a particle by a centre of force is discussed from a wave packet point of view, and in Chapter 7 such important ideas as the scattering matrix are introduced via the wave packet treatment.

So far as possible the notation of Volume 1 is followed. For example, operators are usually shown in Gill Sans type. One exception to this is the set of multiplicative operators such as the potential, which may be denoted by V or V, and the unit operator, which may be denoted by I or 1; normally these will be shown in the second notation, as is more usual. The other exceptions are the Möller operators $\Omega^{\pm}$, and the symmetrizer $\mathscr{S}$, where the use of Gill Sans would involve a departure from normal usage.

The arguments have been spelt out in rather more detail than usual. This is because the book is written primarily for a particular kind of reader; for example, the chemical physicist who is not necessarily a professional theoretician and who needs a gentle introduction to the rigours of present-day scattering theory. In fact, the underlying principles and methods used will all be familiar, but the arguments required are long and essentially mathematical (even if the mathematics is not in itself too severe) and it is in these arguments that the inexperienced reader will fall by the wayside unless the details are given at some length.

It is a pleasure to acknowledge the help given by Professor McWeeny in the writing of this book, the helpful comments by postgraduate students and colleagues in the Department of Mathematics of the University of Nottingham, Professor Seaton for advice and discussion, and Professor Burke for a copy of his notes on atomic and molecular collisions. Finally I would like to thank Mrs. Clough and Mrs. Page for their assistance in typing the manuscript.

J. E. G. Farina

INTRODUCTION

THE International Encyclopedia of Physical Chemistry and Chemical Physics is a comprehensive and modern account of all aspects of the domain of science between chemistry and physics, and is written primarily for the graduate and research worker. The Editors-in-Chief have grouped the subject-matter in some twenty groups (General Topics), each having its own editor. The complete work consists of about one hundred volumes, each volume being restricted to around two hundred pages and having a large measure of independence. Particular importance has been given to the exposition of the fundamental bases of each topic and to the development of the theoretical aspects; experimental details of an essentially practical nature are not emphasized although the theoretical background of techniques and procedures is fully developed.

The Encyclopedia is written throughout in English and the recommendations of the International Union of Pure and Applied Chemistry on notation and cognate matters in physical chemistry are adopted. Abbreviations for names of journals are in accordance with *The World List of Scientific Periodicals*.

CHAPTER 1

SCATTERING OF A PARTICLE BY A CENTRE OF FORCE

1.1. Nature of the scattering problem, cross-sections

Many scattering experiments carried out in laboratories consist of the scattering of electrons by atoms or molecules. The target particles may make up a thin foil or gas. A stream of particles is aimed at the target particle, their velocities usually being as uniform as the experimentalist can attain, and the number of particles emerging in a given direction per unit time is measured. Target atoms or molecules are much more massive than electrons, and in most experiments their velocities before and after the collision will be small compared with the velocity of the incident electron. As a first approximation to the scattering we can take the target particles as infinitely massive and at rest. If we further neglect the structure of the target particles, the problem reduces to the scattering of a stream of incident particles by fixed centres of force. Under certain experimental conditions (and their attainment usually requires skill and care on the part of the experimentalist) it can be assumed that the scattering of an incident particle takes place from only one target particle; that is to say, there are no multiple collisions. Under these circumstances, if there are "n" target particles, the number of particles scattered per unit time in any one direction is n times the corresponding number for scattering of the same stream of particles by only one fixed centre of force. It is this simplified problem, viz. the scattering of a stream of particles of uniform velocity by a fixed centre of force, which will occupy our attention in this chapter.

Let us now consider the scattering of a stream of particles by a fixed centre of force, which we will take as the origin O, and make more precise the ideas introduced above. We take spherical polar coordinates with Oz as polar axis, where Oz is parallel to the direction of the incident beam (see Fig. 1.1). The *flux* I of the incident beam is defined as the number of particles crossing a unit area perpendicular to the direction of the beam in unit time. Given the flux, and type of particle, the nature of the incident beam of particles is determined completely.

In order to define a measure of the scattering, we must make more precise the phrase "number of particles emerging in a given direction per unit time". To do this we consider the solid angle $d\omega$ whose vertex is O and consisting of directions whose angular coordinates lie in the ranges θ, $\theta+d\theta$ and φ, $\varphi+d\varphi$ (see Fig. 1.1). We define the flux $I(\theta, \varphi)d\omega$ of particles into $d\omega$ as the number of particles appearing in $d\omega$ per unit time; this is clearly proportional to the incident flux I and the solid angle $d\omega$. The constant of proportionality depends on θ and φ, and

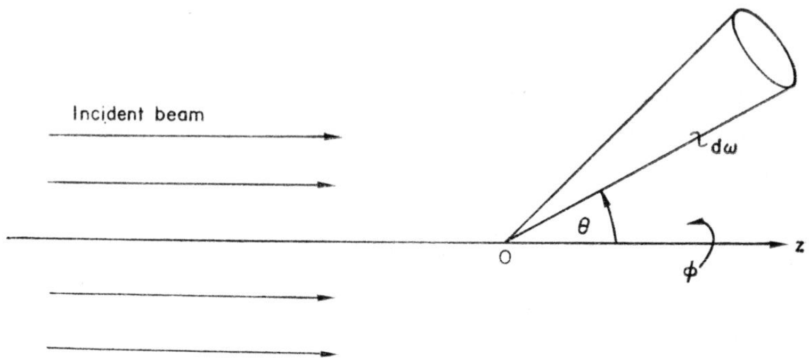

FIG. 1.1. Scattering by a centre of force O.

provides a measure of the strength of scattering in the θ, φ direction; it is denoted by $\sigma(\theta, \varphi)$ and called the *differential cross-section*, so

$$\sigma(\theta, \varphi) = \frac{I(\theta, \varphi)}{I}. \tag{1.1.1}$$

The *total cross-section* σ is defined by

$$\boxed{\sigma = \int \sigma(\theta, \varphi)\, d\omega = \int_0^{\pi} \sin\theta\, d\theta \int_0^{2\pi} d\varphi\, \sigma(\theta, \varphi).} \tag{1.1.2}$$

$I\sigma$ is thus the number of particles scattered per unit time, so that σ is a measure of the total scattering by the centre of force. Both σ and $\sigma(\theta, \varphi)$ will depend on the speed (or energy) of the incident particles, and will also vary for different incident particles and for different forces.

Since $I(\theta, \varphi)$ is a number of particles per unit time, and I is a number of particles per unit time per unit area, (1.1.1) shows that $\sigma(\theta, \varphi)$ has the dimensions of area. From (1.1.2) it follows that σ also has the dimensions of area. Classically the picture will be as shown in Fig. 1.2, if we assume spherical symmetry. Figure 1.2(b) shows the picture as seen by an observer looking along Oz from the positive direction. The number of particles scattered into the solid angle $d\omega = \sin\theta\, d\theta\, d\varphi$ will be the number of particles whose initial trajectories have perpendicular distances from O lying between some distance R and some distance $R+dR$, and with azimuthal angle lying between φ and $\varphi+d\varphi$. In general, since we have taken $d\theta$ as positive, dR will be negative, since the angle of scattering θ will decrease with increasing R. The number of particles $I(\theta, \varphi)\, d\omega$ entering the solid angle $d\omega$ per unit time will be the

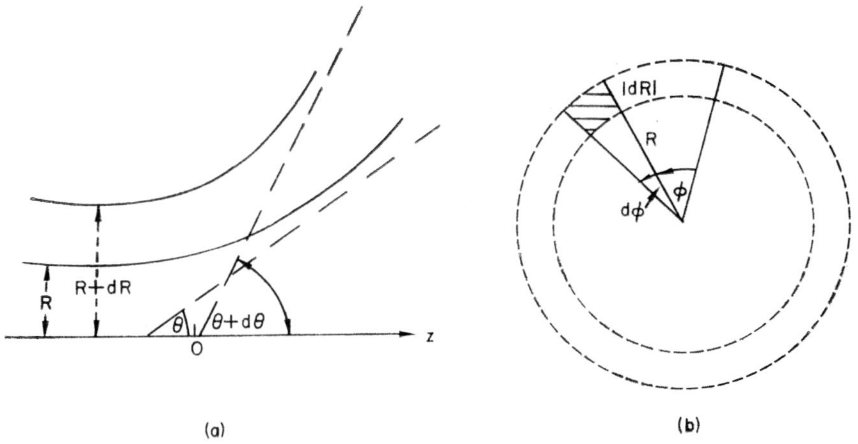

FIG. 1.2. Classical scattering.

number of particles crossing the shaded area on Fig. 1.2(b) per unit time; in other words, $IR|dR|d\varphi$. Hence by (1.1.1),

$$\sigma(\theta, \varphi)\, d\omega = \frac{IR|dR|d\varphi}{I},$$

$$\therefore\ \sigma(\theta, \varphi) \sin\theta\, d\theta\, d\varphi = R|dR|d\varphi,$$

$$\therefore\ \sigma(\theta, \varphi) = R \operatorname{cosec}\theta \left|\frac{dR}{d\theta}\right|, \tag{1.1.3}$$

confirming that $\sigma(\theta, \varphi)$ has the dimensions of area. The total cross-section σ is obtained by substitution for $\sigma(\theta, \varphi)$ from (1.1.3) into (1.1.2), so

$$\sigma = \int\int R|dR|\, d\varphi.$$

$$\therefore\ \sigma = \int 2\pi\, R|dR|. \tag{1.1.4}$$

The limits on R in (1.1.4) will usually be from zero to some maximum distance R_0 beyond which the particles pass without deviation, and so

$\sigma = \pi R_0^2$. In other words, σ is just the area of the circular cross-section through which particles pass if they are to be scattered, which is the effective cross-section.

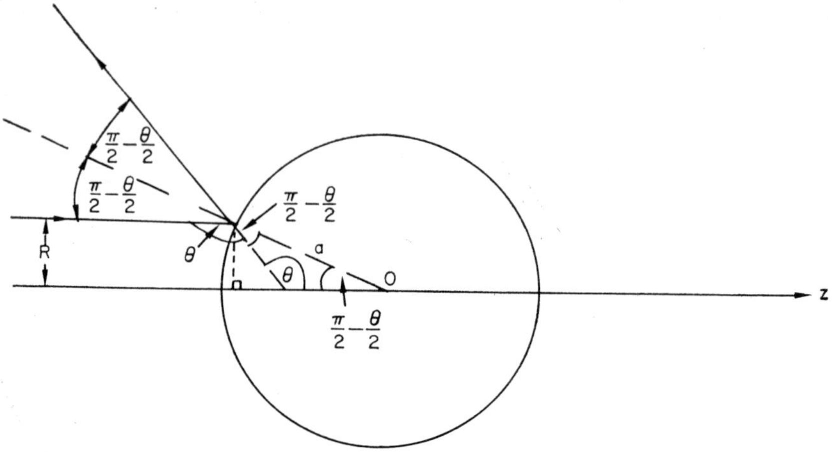

FIG. 1.3. Classical scattering by a hard sphere.

EXAMPLE 1.1. *Classical scattering by a hard sphere.* As an example, let us evaluate the differential and total cross-sections for the classical scattering of a particle by a hard, smooth, perfectly elastic sphere of radius a. From Fig. 1.3 we have

$$R = a \sin(\tfrac{1}{2}\pi - \tfrac{1}{2}\theta) = a \cos \tfrac{1}{2}\theta; \tag{1.1.5}$$

$$\therefore \frac{dR}{d\theta} = -\tfrac{1}{2}a \sin \tfrac{1}{2}\theta. \tag{1.1.6}$$

Application of (1.1.3) gives

$$\sigma(\theta, \varphi) = a \cos \tfrac{1}{2}\theta . \operatorname{cosec} \theta . \tfrac{1}{2}a \sin \tfrac{1}{2}\theta = \tfrac{1}{4}a^2. \tag{1.1.7}$$

Hence by (1.1.2),

$$\sigma = \int \tfrac{1}{4}a^2\, d\omega = \tfrac{1}{4}a^2 . 4\pi = \pi a^2. \tag{1.1.8}$$

We notice that the differential cross-section is constant, so that, for example, backward scattering ($\theta = 180°$) is just as strong as forward scattering ($\theta = 0°$). The total cross-section is πa^2, as we should expect.

1.2. Quantum mechanical formulation

Quantum mechanically we have to formulate the problem so as to describe an incident beam together with outgoing particles (see Fig. 1.4). The wave function describing an incident beam normalized to one particle per unit volume and moving parallel to Oz with momentum

$\mu v = \hbar k$ where μ is the mass of each particle, v its velocity, and k its wave number, is the plane wave

$$\Phi_{\mathbf{k}}(\mathbf{r}) = \exp(ikz) = \exp(i\mathbf{k}\cdot\mathbf{r}), \tag{1.2.1}$$

where $\mathbf{k}$ is the vector with Cartesian components $(0, 0, k)$ (see, for example, Volume 1, Section 2.4).

The wave function Φ_s describing the scattered particles is what is called a "spherical wave". It has the form

$$\Phi_s = f(\theta, \varphi)\frac{\exp(ikr)}{r}. \tag{1.2.2}$$

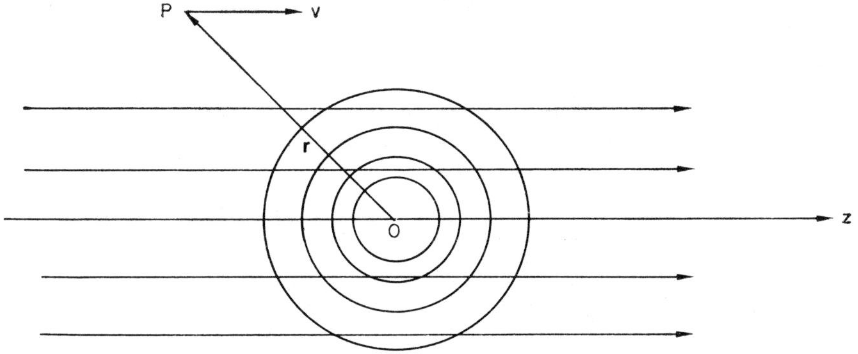

FIG. 1.4. Scattering of the incident beam.

We note that for large values of r, Φ_s satisfies the time-independent Schrödinger equation for the free particle, viz.

$$-\frac{\hbar^2}{2\mu}\nabla^2\Phi_s = E\,\Phi_s \tag{1.2.3}$$

where E is the energy of the free particle, i.e.

$$E = \tfrac{1}{2}\mu v^2 = \hbar^2 k^2/(2\mu). \tag{1.2.4}$$

For, taking spherical polar coordinates (r, θ, φ) with Oz as polar axis, the Laplacian operator has the form (see also Volume 2, Section 2.4)

$$\begin{aligned}\nabla^2 &= \frac{\partial^2}{\partial r^2}+\frac{2}{r}\frac{\partial}{\partial r}+\frac{1}{r^2}\frac{\partial^2}{\partial\theta^2}+\frac{\cot\theta}{r^2}\frac{\partial}{\partial\theta}+\frac{1}{r^2\sin^2\theta}\frac{\partial^2}{\partial\varphi^2}\\ &= \frac{\partial^2}{\partial r^2}+\frac{2}{r}\frac{\partial}{\partial r}+\frac{1}{r^2}\Lambda, \text{ say.}\end{aligned} \tag{1.2.5}$$

Hence from (1.2.2), as $r \to \infty$,

$$-\frac{\hbar^2}{2\mu} \nabla^2 \Phi_s = \frac{\hbar^2 k^2}{2\mu} \Phi_s + \text{terms tending to zero faster than } \frac{1}{r}.$$

$$= E\,\Phi_s + \text{terms tending to zero faster than } \frac{1}{r},$$

and so Φ_s satisfies (1.2.3) as $r \to \infty$.

Now the vector current $\mathbf{J}$ arising from Φ_s is (Volume 1, Section 1.3)

$$\mathbf{J} = -\frac{i\hbar}{2\mu} (\Phi_s^* \nabla\Phi_s - \Phi_s \nabla\Phi_s^*), \tag{1.2.6}$$

where the star indicates the complex conjugate. The radial component J_r of $\mathbf{J}$ (i.e. the component of $\mathbf{J}$ in the direction outgoing from O) is therefore given by

$$J_r = -\frac{i\hbar}{2\mu}\left[f^* \frac{\exp(-ikr)}{r} \frac{\partial}{\partial r}\left\{f \frac{\exp(ikr)}{r}\right\} - f \frac{\exp(ikr)}{r} \frac{\partial}{\partial r}\left\{f^* \frac{\exp(-ikr)}{r}\right\}\right].$$

If we carry out the differentiations and simplify we find that

$$J_r = \frac{|f|^2 \hbar k}{\mu r^2} = \frac{|f|^2 \mu v}{\mu r^2} = \frac{v|f|^2}{r^2}. \tag{1.2.7}$$

The flux $I(\theta, \varphi)\,d\omega$ in the solid angle $d\omega$ is the number of particles emerging in $d\omega$ per unit time, and is therefore J_r times the cross-sectional area $r^2\,d\omega$ of $d\omega$ perpendicular to the radius, so

$$I(\theta, \varphi)\,d\omega = J_r\, r^2\, d\omega = \frac{v|f|^2}{r^2} r^2\, d\omega = v|f|^2\, d\omega. \tag{1.2.8}$$

Since the incident beam is normalized to one particle per unit volume, the incident flux I is v. Hence (1.1.1) and (1.2.8) give

$$\boxed{\sigma(\theta, \varphi) = |f(\theta, \varphi)|^2.} \tag{1.2.9}$$

The wave function Ψ which describes the scattering process must satisfy Schrödinger's equation for the motion of a particle in the presence of a potential $V = V(\mathbf{r})$, viz.

$$\left\{-\frac{\hbar^2}{2\mu}\nabla^2 + V\right\}\Psi = E\,\Psi. \tag{1.2.10}$$

At large distances the potential V will be negligible, and since it must describe both freely moving incident particles and freely moving scattered particles, we expect it to become like $\Phi_k + \Phi_s$. Thus,

$$\Psi \underset{r\to\infty}{\sim} \Phi_k + \Phi_s$$

or, in full,

$$\Psi(\mathbf{r}) \underset{r\to\infty}{\sim} \exp(ikz) + f(\theta, \varphi)\frac{\exp(ikr)}{r}. \tag{1.2.11}$$

Equation (1.2.11) is known as a "boundary condition" (Churchill, 1963). All wave functions are further restricted to being continuous at all points, and in particular at the origin; Ψ must be "regular" at the origin. We may sum up these conclusions as follows:

> In order to obtain the differential cross-section, the wave equation (1.2.10) must be solved for a wave function Ψ, regular at the origin, subject to the boundary condition (1.2.11). The differential cross-section is then obtained from (1.2.9).

The quantity $f(\theta, \varphi)$ is known as the *scattering amplitude*. This is the factor which determines the relevant physical quantities, the cross-sections.

In order to solve the scattering problem, we must first find a solution —a question which will be examined in subsequent sections. Having thus obtained a scattering amplitude $f(\theta, \varphi)$, we must be certain that it is the only such one, for otherwise our problem would not have a unique answer, as is required by physics. That this is indeed so follows by the following argument:

Firstly, if we put $U = 2\mu V/\hbar^2$ and remember that $E = \hbar^2k^2/2\mu$, (1.2.10) can be rewritten

$$(\nabla^2 + k^2)\Psi = U\Psi. \tag{1.2.12}$$

If Ψ_1 and Ψ_2 satisfy (1.2.12) together with the boundary condition (1.2.11), i.e.

$$\Psi_n(\mathbf{r}) \underset{r\to\infty}{\sim} \exp(ikz)+f_n(\theta, \varphi)\frac{\exp(ikr)}{r}, \quad (n = 1, 2), \tag{1.2.13}$$

then $\Psi = \Psi_1 - \Psi_2$ satisfies (1.2.12) together with the boundary condition

$$\Psi(\mathbf{r}) \underset{r\to\infty}{\sim} (f_1-f_2)\frac{\exp(ikr)}{r} = f\frac{\exp(ikr)}{r}, \text{ say.} \tag{1.2.14}$$

The current $\mathbf{J}$ is given by (1.2.6) with Φ_s replaced by Ψ, and so if we take the divergence of $\mathbf{J}$ we easily see that

$$\text{div } \mathbf{J} = -\frac{i\hbar}{2\mu}(\Psi^* \nabla^2 \Psi - \Psi \nabla^2 \Psi^*). \tag{1.2.15}$$

The total Hamiltonian H is given by

$$\mathsf{H} = -\frac{\hbar^2}{2\mu}\nabla^2 + V. \tag{1.2.16}$$

Also, since H is real and $\mathsf{H}\Psi = E\Psi$, (1.2.15) and (1.2.16) give

$$\text{div } \mathbf{J} = \frac{i}{\hbar}[\Psi^* \mathsf{H}\Psi - \Psi(\mathsf{H}\Psi)^*] = 0. \tag{1.2.17}$$

Now let V be the region enclosed by a sphere S of radius R and centre the origin. Since Ψ is regular at the origin and div $\mathbf{J} = 0$ in V we may apply the divergence theorem to obtain (using J_r for the radial component)

$$\int_S J_r\, dS = \int_S \mathbf{J}\cdot d\mathbf{S} = \int_v \text{div } \mathbf{J}\, dV = 0. \tag{1.2.18}$$

By (1.2.7), $J_r = v|f|^2/r^2 = v|f|^2/R^2$ on S, hence if we substitute for J_r into (1.2.18), divide by v, and put $dS = R^2\, d\omega$ where $d\omega$ is the solid angle subtended by dS at O, we have

$$\int_S |f|^2\, d\omega = 0. \tag{1.2.19}$$

Since $|f|$ is continuous and $\geqq 0$, this can only be satisfied if $|f| = 0$, and so $f_1 = f_2$, proving the uniqueness of the solution.

1.3. Some important differential equations

We will make a diversion from the main scattering problem at this stage in order to remind ourselves of some well-known differential equations and their solutions, whose importance arises from their frequent occurrence in physical problems and, in particular, in scattering problems. For further details the reader is referred to Sneddon (1961) or to the short review in Volume 1, Appendix 3.

Legendre's equation

The first equation required is Legendre's equation of order ν. This is

$$(1-x^2)\frac{d^2y}{dx^2}-2x\frac{dy}{dx}+\nu(\nu+1)y = 0. \tag{1.3.1}$$

It is easy to see that the functions

$$u_\nu(x) = 1-\frac{\nu(\nu+1)}{2!}x^2+\frac{(\nu-2)\nu\,(\nu+1)\,(\nu+3)}{4!}x^4-\dots \tag{1.3.2}$$

and

$$v_\nu(x) = x-\frac{(\nu-1)\,(\nu+2)}{3!}x^3+\frac{(\nu-3)\,(\nu-1)\,(\nu+2)\,(\nu+4)}{5!}x^5-\dots \tag{1.3.3}$$

are solutions of (1.3.1), either by direct substitution of u_ν and v_ν for y, or by solving (1.3.1) in series.

One should note that u_ν and v_ν are "linearly independent"; that is to say, one is not a multiple of the other. Equation (1.3.1) is a "linear" one, in the sense that if y_1 and y_2 are solutions, so is Ay_1+By_2, where A and B are arbitrary constants. In particular,

$$y = Au_\nu+Bv_\nu \tag{1.3.4}$$

is a solution of (1.3.1), and since it contains two arbitrary constants and the differential equation is of the second order, it must be the general solution.

We note that if ν is even, the first series must terminate so that u_ν is a polynomial, while if ν is odd, v_ν is a polynomial. The *Legendre polynomials* $P_n(x)$ $(n = 0, 1, 2\dots)$ may be defined by

$$P_n(x) = \begin{cases} u_n(x)/u_n(1), & n \text{ even;} \\ v_n(x)/v_n(1), & n \text{ odd.} \end{cases} \tag{1.3.5}$$

The Legendre polynomials form an *orthogonal set* on the interval

$-1 \leqq x \leqq +1$, by which we mean that

$$\int_{-1}^{+1} P_m(x)\, P_n(x)\, dx = 0, \quad (m \neq n). \tag{1.3.6}$$

In fact it can be proved that

$$\int_{-1}^{+1} P_m(x)\, P_n(x)\, dx = \frac{2}{2n+1}\, \delta_{mn} \tag{1.3.7}$$

where δ_{mn} is the Kronecker-δ defined by

$$\delta_{mn} = \begin{cases} 0, & m \neq n, \\ 1, & m = n. \end{cases} \tag{1.3.8}$$

The properties of complete orthogonal sets are discussed in the standard textbooks (e.g. Churchill, 1963) and an elementary account is available in Volume 1, Chapter 3.

The Legendre polynomials form a *complete* set for the interval $-1 \leqq x \leqq +1$, in the sense that if f is a smooth function (i.e. a differentiable function), then

$$f(x) = \sum_{n=0}^{\infty} A_n\, P_n(x), \quad -1 \leqq x \leqq 1, \tag{1.3.9}$$

where the A_n may be obtained by multiplying (1.3.9) by P_m and integrating with respect to x from $x = -1$ to $x = +1$. On use of (1.3.7) we obtain

$$\int_{-1}^{+1} P_m(x)\, f(x)\, dx = \sum_{n=0}^{\infty} A_n \frac{2}{2n+1}\, \delta_{mn} = A_m \frac{2}{2m+1};$$

hence

$$A_n = \frac{2n+1}{2} \int_{-1}^{+1} P_n(x)\, f(x)\, dx. \tag{1.3.10}$$

According to (1.2.5)

$$\nabla^2 = \frac{\partial^2}{\partial r^2} + \frac{2}{r}\frac{\partial}{\partial r} + \frac{1}{r^2}\Lambda \tag{1.3.11}$$

where

$$\Lambda = \frac{\partial^2}{\partial \theta^2} + \cot\theta \frac{\partial}{\partial \theta} + \frac{1}{\sin^2\theta}\frac{\partial^2}{\partial \varphi^2}, \tag{1.3.12}$$

hence if $\cos\theta = x$ and $l = 0, 1, 2 \ldots$

$$\begin{aligned}\Lambda P_l(\cos\theta) &= \left[\frac{\partial^2}{\partial\theta^2}+\cot\theta\,\frac{\partial}{\partial\theta}\right] P_l(\cos\theta)\\ &= \frac{1}{\sin\theta}\frac{\partial}{\partial\theta}\sin\theta\,\frac{\partial}{\partial\theta}P_l(\cos\theta)\\ &= \frac{1}{\sin\theta}\frac{\partial}{\partial\theta}\sin^2\theta\,\frac{\partial}{\sin\theta\,\partial\theta}P_l(\cos\theta)\\ &= \frac{\partial}{\partial x}(1-x^2)\frac{\partial}{\partial x}P_l(x)\\ &= (1-x^2)\,P_l''(x)-2x\,P_l'(x).\end{aligned} \tag{1.3.13}$$

If we remember that $P_l(x)$ satisfies (1.3.1) with $\nu = l$, (1.3.13) gives

$$\Lambda P_l(\cos\theta) = -l(l+1)\,P_l(\cos\theta). \tag{1.3.14}$$

Now it is shown in books on quantum mechanics (e.g. Volume 2, Section 2.4) that if $\mathbf{L}$ is the orbital angular momentum operator for a single particle, then

$$\mathbf{L}^2 = -\hbar^2\,\Lambda. \tag{1.3.15}$$

Hence (1.3.14) can be rewritten

$$\mathbf{L}^2 P_l = \hbar^2\, l(l+1)\,P_l \tag{1.3.16}$$

showing that P_l is an eigenfunction of $\mathbf{L}^2$ corresponding to total angular momentum $\hbar^2\,l(l+1)$. The bound states corresponding to $l = 0, 1, 2 \ldots$ are usually referred to as $s, p, d \ldots$ states, and this is a nomenclature which is also adopted in scattering theory.

Bessel's equation

Another equation of frequent occurrence is *Bessel's equation of order* ν, which is

$$\frac{d^2y}{dx^2}+\frac{1}{x}\frac{dy}{dx}+\left(1-\frac{\nu^2}{x^2}\right)y = 0. \tag{1.3.17}$$

This may also be solved in series, and if ν is not an integer the general solution of (1.3.17) is found to be

$$y = AJ_\nu(x)+BJ_{-\nu}(x) \tag{1.3.18}$$

where

$$J_\nu(x) = \sum_{n=0}^{\infty}\frac{(-1)^n\,(x/2)^{\nu+2n}}{n!\,\Gamma(n+\nu+1)} \tag{1.3.19}$$

and Γ denotes the gamma function (Sneddon, 1961). We immediately notice that

$$J_\nu(x) \underset{x\to 0}{\sim} \frac{x^\nu}{2^\nu \Gamma(\nu+1)}. \tag{1.3.20}$$

Thus $J_\nu \to 0$ as $x \to 0$ and is defined at the origin, where it equals zero, whilst $J_{-\nu} \to \infty$ as $x \to 0$ and is not defined at the origin. For these reasons we call J_ν the regular solution of (1.3.17) and $J_{-\nu}$ the irregular solution. If the solution (1.3.18) is to represent some physical quantity at the origin, it is clear that we must choose B to be zero.

In the next section we shall come across the differential equation

$$\left[\frac{d^2}{dr^2} - \frac{l(l+1)}{r^2} + k^2\right] y = 0 \quad (l = 0, 1, 2, \ldots). \tag{1.3.21}$$

The solution of this equation may be obtained in terms of Bessel functions, if firstly we divide by k^2 and put $x = kr$, so that (1.3.21) becomes

$$\left[\frac{d^2}{dx^2} - \frac{l(l+1)}{x^2} + 1\right] y = 0. \tag{1.3.22}$$

A solution of (1.3.22) is $x^{\frac{1}{2}} J_{l+\frac{1}{2}}(x)$, for

$$\left[\frac{d^2}{dx^2} - \frac{l(l+1)}{x^2} + 1\right] x^{\frac{1}{2}} J_{l+\frac{1}{2}}(x)$$
$$= x^{\frac{1}{2}}\left[J''_{l+\frac{1}{2}}(x) + \frac{1}{x} J'_{l+\frac{1}{2}}(x) + \left\{1 - \frac{(l+\frac{1}{2})^2}{x^2}\right\} J_{l+\frac{1}{2}}(x)\right] = 0 \tag{1.3.23}$$

since $J_{l+\frac{1}{2}}$ satisfies Bessel's equation (1.3.17) with $\nu = l+\frac{1}{2}$.

The *spherical Bessel function* (of order l), $j_l(x)$, is defined by

$$j_l(x) = \left(\frac{\pi}{2x}\right)^{\frac{1}{2}} J_{l+\frac{1}{2}}(x). \tag{1.3.24}$$

Thus $xj_l(x) = (\pi/2)^{\frac{1}{2}} x^{\frac{1}{2}} J_{l+\frac{1}{2}}(x)$, and so it follows that $xj_l(x)$ is a solution of (1.3.22), and hence $krj_l(kr)$ is a solution of (1.3.21).

The other solution of (1.3.22) is $x^{\frac{1}{2}} J_{-l-\frac{1}{2}}(x)$, and if $l \neq 0$, (1.3.20) shows that this is infinite at the origin, so that this is the *irregular* solution. We define the *spherical Neumann function* $n_l(x)$ by

$$n_l(x) = (-1)^{l+1} \left(\frac{\pi}{2x}\right)^{\frac{1}{2}} J_{-l-\frac{1}{2}}(x), \tag{1.3.25}$$

so that $xn_l(x)$ is an irregular solution of (1.3.22) and $krn_l(kr)$ is an irregular solution of (1.3.21). All other irregular solutions must be multiples of these.

The functions j_l and n_l may be proved to have the following properties (Watson, 1958):

$$j_l(x) \underset{x\to 0}{\sim} \frac{x^l}{1.3.5\ldots(2l+1)}; \tag{1.3.26}$$

$$n_l(x) \underset{x\to 0}{\sim} \frac{(-1).1.1.3.5\ldots(2l-1)}{x^{l+1}}; \tag{1.3.27}$$

$$j_l(x) \underset{x\to \infty}{\sim} \frac{1}{x}\cos\left[x-\tfrac{1}{2}(l+1)\pi\right]; \tag{1.3.28}$$

$$n_l(x) \underset{x\to \infty}{\sim} \frac{1}{x}\sin\left[x-\tfrac{1}{2}(l+1)\pi\right]. \tag{1.3.29}$$

From (1.3.26) we see that $krj_l(kr) = 0$ when $r = 0$, whereas from (1.3.27) $krn_l(kr) \neq 0$ when $r = 0$, a fact of importance for the scattering problem.

When $l = 0$, (1.3.22) becomes just the equation of simple harmonic motion with solutions $\sin x$ and $\cos x$; in fact

$$j_0(x) = \frac{\sin x}{x}, \quad n_0(x) = -\frac{\cos x}{x}, \tag{1.3.30}$$

and equations (1.3.26)–(1.3.29) are obvious in this case.

The hypergeometric equation

Another equation of importance is

$$x\frac{d^2y}{dx^2}+(b-x)\frac{dy}{dx}-ay = 0. \tag{1.3.31}$$

If we try a solution of the form

$$y = \sum_{n=0}^{\infty} a_n x^n, \quad (a_0 \neq 0), \tag{1.3.32}$$

we find that

$$a_n = \frac{(n-1+a)a_{n-1}}{n(n-1+b)}, \quad (n \geqq 1), \tag{1.3.33}$$

and hence that

$$a_n = \frac{(n-1+a)(n-2+a)\ldots a}{n!\,(n-1+b)(n-2+b)\ldots b}\,a_0. \tag{1.3.34}$$

The gamma function (Sneddon, 1961) has the property that $x\Gamma(x) = \Gamma(x+1)$, so (1.3.34) can be rewritten

$$a_n = \frac{\{\Gamma(n+a)/\Gamma(a)\}\,a_0}{\{\Gamma(n+b)/\Gamma(b)\}\,n!}. \tag{1.3.35}$$

If we take $a_0 = 1$, the solution (1.3.32) takes the form

$$y = \sum_{n=0}^{\infty} \frac{\{\Gamma(n+a)/\Gamma(a)\}}{\{\Gamma(n+b)/\Gamma(b)\}\, n!}\, x^n. \tag{1.3.36}$$

The expression (1.3.36) is known as the "confluent hypergeometric function" and denoted by ${}_1F_1(a, b; x)$, while the differential equation (1.3.31), of which it is the regular solution, is known as the "confluent hypergeometric equation" (Sneddon, 1961, p. 34). Equation (1.3.31) is also sometimes referred to as "Laplace's equation".

1.4. Partial wave analysis

In this section we return to the main problem. We have seen that to obtain the differential and total cross-sections, we have to find a solution of the Schrödinger equation (1.2.10) which, for physical reasons, must be regular at the origin and satisfy the boundary condition (1.2.11). Once we have found such a solution the scattering amplitude, and hence the differential cross-section, will be determined uniquely. We shall now proceed to obtain such a solution, subject to the assumption that the problem has spherical symmetry, i.e. the potential U is a function only of r, the distance from the origin. Thus from (1.2.12) we seek a solution of

$$(\nabla^2 + k^2 - U)\Psi = 0 \tag{1.4.1}$$

where $U = U(r)$.

It is reasonable to look in the first place for a solution in the form $\Psi = \Psi(r, \theta)$; that is, to assume independence of φ (i.e. axial symmetry around the beam). In spherical polar coordinates θ runs from 0 to π, and hence θ is a function of $\cos\theta$, since given $\cos\theta$, θ is determined uniquely in this interval; hence we can write $\Psi = \Psi(r, \cos\theta)$. Now the Legendre polynomials form a complete set, and so

$$\Psi = \Psi(r, \cos\theta) = \sum_{l=0}^{\infty} A_l(r)\, P_l(\cos\theta). \tag{1.4.2}$$

The coefficients A_l depend, of course, on r, but are otherwise unspecified. It will prove more convenient for us to rewrite (1.4.2) in the form

$$\Psi = \sum_{l=0}^{\infty} C_l \frac{1}{r} F_l(r)\, P_l(\cos\theta) \tag{1.4.3}$$

where $C_0, C_1, C_2, \ldots$ are as yet undetermined numbers, and $F_0, F_1, F_2, \ldots$ undetermined functions. We now substitute for Ψ from (1.4.3) into (1.4.1) and use (1.3.11) to get

$$\left[\frac{\partial^2}{\partial r^2}+\frac{2}{r}\frac{\partial}{\partial r}+\frac{1}{r^2}\Lambda+k^2-U\right]\sum_{l=0}^{\infty} C_l \frac{1}{r} F_l(r)\, P_l(\cos\theta) = 0. \tag{1.4.4}$$

If we further assume that the infinite sum on the left-hand side of (1.4.4) can be differentiated term by term we can use (1.3.14) to obtain

$$\sum_{l=0}^{\infty} C_l \left[\frac{\partial^2}{\partial r^2}+\frac{2}{r}\frac{\partial}{\partial r}-\frac{l(l+1)}{r^2}+k^2-U\right]\frac{1}{r} F_l(r)\, P_l(\cos\theta) = 0. \tag{1.4.5}$$

Equation (1.4.5) is true for all θ, and so the coefficient of each P_l can be equated to zero. Dividing each such equation by C_l, carrying out the differentiations, and multiplying by r we find that

$$\boxed{\left[\frac{d^2}{dr^2}-\frac{l(l+1)}{r^2}-U(r)+k^2\right]F_l(r) = 0.} \tag{1.4.6}$$

Equation (1.4.6) is known as the *radial equation*, since it involves only the radial coordinate r. In principle we have to solve it for all l; then (1.4.3) gives us a solution for all C_l such that the series converges and the interchange of differentiation and summation is justified. Now if U vanishes for $r > a$ say, or tends to zero sufficiently fast, (1.4.6) may be approximated by (for large r)

$$\left[\frac{d^2}{dr^2}+k^2\right]F_l(r) = 0. \tag{1.4.7}$$

Thus for large r we expect our solutions to be essentially sinusoidal; in other words, we expect that

$$F_l(r) \underset{r\to\infty}{\sim} A_l \sin\left(kr-\tfrac{1}{2}l\pi+\eta_l\right) \tag{1.4.8}$$

where A and η_l are constants. Furthermore, we require that Ψ is regular at the origin, and so from (1.4.3) $F_l(0) = 0$. Since the solution is usually a linear combination of two functions, only one of which vanishes when $r = 0$, this requirement determines F_l to within a multiplicative constant, and so η_l is thus determined but A_l is still indeterminate. If we compare (1.4.6) with (1.3.21), which is the same equation but with the interaction U absent, and remember that the regular solution $krj_l(kr)$ of (1.3.21) has the asymptotic form [from (1.3.28)]

$$krj_l(kr) \underset{r\to\infty}{\sim} \sin\left(kr-\tfrac{1}{2}l\pi\right) \tag{1.4.9}$$

we see that, if $A_l = 1$, the effect of the potential U is to introduce a shift in the phase of the lth scattered wave of amount η_l; for this reason η_l is known as the lth *phase shift*. Since the C_l in (1.4.3) are still un specified we are at liberty to choose $A_l = 1$, so that (1.4.8) becomes

$$\boxed{F_l(r) \underset{r\to\infty}{\sim} \sin\left(kr-\tfrac{1}{2}l\pi+\eta_l\right).} \tag{1.4.10}$$

The choice $A_l = 1$ is called the *normalization* of the F_l. This is analogous to the normalization of a bound state by appropriate choice of a multiplicative constant, but is *not* the same, since the function F_l is clearly not square integrable. The function F_l is usually known as the lth *partial wave*, and in particular F_0, F_1, F_2 are known as s-waves, p-waves, d-waves, etc., since as we saw in Section 1.3 they correspond to wave functions of total angular momentum $l = 0$, $l = 1$, $l = 2$, etc.

Let us now consider the special case when $U = 0$ so that (1.4.1) becomes

$$(\nabla^2+k^2)\Psi = 0. \tag{1.4.11}$$

The above procedure tells us that the solution now has the form

$$\Psi = \Psi_0 = \sum_{l=0}^{\infty} C_l^0 \frac{1}{r} F_l^0(r)\, P_l(\cos\theta) \tag{1.4.12}$$

where F_l^0 satisfies the radial equation (1.4.6) with $U = 0$, viz.

$$\left[\frac{d^2}{dr^2}-\frac{l(l+1)}{r^2}+k^2\right]F_l(r) = 0, \tag{1.4.13}$$

together with [from (1.4.10)], the boundary condition

$$F_l^0(r) \underset{r\to\infty}{\sim} \sin\left(kr-\tfrac{1}{2}l\pi+\eta_l^0\right). \tag{1.4.14}$$

Now (1.4.13) is identical with (1.3.21), and if Ψ_0 is regular at the origin $F_l^0(0) = 0$, which means that F_l^0 is proportional to $krj_l(kr)$. The condition (1.3.28) shows that

$$krj_l(kr) \underset{r\to\infty}{\sim} \sin\left(kr-\tfrac{1}{2}l\pi\right) \tag{1.4.15}$$

and comparing this with (1.4.14) we see that the constant of proportionality is unity and $\eta_l^0 = 0$. Hence if Ψ_0 is a regular solution of

(1.4.11), (1.4.12) gives us

$$\Psi_0 = \sum_{l=0}^{\infty} C_l^0\, kj_l(kr)\, P_l(\cos\theta). \tag{1.4.16}$$

In particular $\exp(ikz) = \exp(ikr\cos\theta)$ is such a solution of (1.4.11), and hence if $t = \cos\theta$, (1.4.16) gives

$$\exp(ikz) = \exp(ikrt) = \sum_{n=0}^{\infty} C_n^0\, kj_n(kr)\, P_n(t). \tag{1.4.17}$$

To determine the C_n^0, we multiply (1.4.17) by $P_l(t)$, integrate from $t = -1$ to $t = +1$, and use the orthogonality property (1.3.7); then on integrating by parts, we have

$$\begin{aligned} C_l^0\, kj_l(kr)\frac{2}{2l+1} &= \int_{-1}^{+1} \exp(ikrt)\, P_l(t)\, dt \\ &= \left[\frac{\exp(ikrt)}{ikr} P_l(t)\right]_{t=-1}^{t=+1} - \int_{-1}^{+1} \frac{\exp(ikr)}{ikr} P_l'(t)\, dt \\ &= \frac{\exp(ikr)-(-1)^l \exp(-ikr)}{ikr} + T \end{aligned} \tag{1.4.18}$$

where by repeated integration by parts we can show that $T \to 0$ as $r \to \infty$ faster than $1/r$, and we have used the fact that $P_l(-1) = (-1)^l$ (Sneddon, 1961). Hence

$$\begin{aligned} \frac{2kj_l(kr)}{2l+1} C_l^0 &= \frac{i^l[\exp\{i(kr-\frac{1}{2}l\pi)\} - \exp\{-i(kr-\frac{1}{2}l\pi)\}]}{ikr} + T \\ &= \frac{2i^l \sin(kr-\frac{1}{2}l\pi)}{kr} + T. \end{aligned} \tag{1.4.19}$$

If we let $r \to \infty$ and substitute the asymptotic form (1.3.28) for $j_l(kr)$, (1.4.19) gives us $C_l^0 = (2l+1)i^l/k$. Hence (1.4.17) gives us

$$\exp(ikz) = \sum_{l=0}^{\infty} (2l+1)i^l\, j_l(kr)\, P_l(\cos\theta). \tag{1.4.20}$$

Equation (1.4.20) is an important result, being the partial wave expansion of a plane wave. Using (1.3.28), (1.4.20) gives us

$$\exp(ikz) \underset{r\to\infty}{\sim} \sum_{l=0}^{\infty} \frac{(2l+1)i^l}{kr} \sin(kr-\tfrac{1}{2}l\pi)\, P_l(\cos\theta). \tag{1.4.21}$$

We now return to the general solution (1.4.3). By (1.4.10), this has asymptotic form

$$\Psi \underset{r\to\infty}{\sim} \sum_{l=0}^{\infty} C_l \frac{1}{r} \sin (kr - \tfrac{1}{2}l\pi + \eta_l)\, P_l(\cos\theta) \tag{1.4.22}$$

and hence if we subtract (1.4.21) from (1.4.22), replace the sines by differences of exponentials multiplied by $1/(2i)$, and rearrange,

$$\begin{aligned}\Psi - \exp(ikz) \underset{r\to\infty}{\sim} & \frac{\exp(ikr)}{r} \sum_{l=0}^{\infty} P_l(\cos\theta) \left[\frac{C_l}{2i} \exp\{i(-\tfrac{1}{2}l\pi + \eta_l)\}\right. \\ & \left. - \frac{(2l+1)i^l \exp(-\tfrac{1}{2}l\pi i)}{2ki}\right] + \frac{\exp(-ikr)}{r} \sum_{l=0}^{\infty} P_l(\cos\theta) \times \\ & \times \left[-\frac{C_l}{2i} \exp\{i(\tfrac{1}{2}l\pi - \eta_l)\} + \frac{(2l+1)i^l \exp(\tfrac{1}{2}l\pi i)}{2ki}\right].\end{aligned} \tag{1.4.23}$$

Now we are seeking a solution which satisfies the boundary condition (1.2.11), and so $\Psi - \exp(ikz)$ must contain only the term in $\exp(ikr)/r$. To ensure this, we *choose* C_l according to

$$C_l = \frac{(2l+1)i^l \exp(i\eta_l)}{k}, \tag{1.4.24}$$

for this causes the term in $\exp(-ikr)/r$ to vanish. Had we chosen C_l to make the first bracket vanish, we would have obtained a solution whose asymptotic form contained incoming waves only. (The reader can easily verify that a wave function $f(\theta, \varphi) \exp(-ikr)/r$ gives rise to an incoming radial current.) This second type of solution is mathematically possible although it does not represent a state normally found in the laboratory; for the moment we shall ignore it, although later we shall see its importance in the further theoretical development of the subject. With the choice (1.4.24) for C_l, we see that the scattering amplitude $f(\theta, \varphi)$, which is the coefficient of $\exp(ikr)/r$ in the asymptotic form (1.4.23), is given by

$$\boxed{f(\theta, \varphi) = f(\theta) = (2ik)^{-1} \sum_{l=0}^{\infty} (2l+1)\,[\exp(2i\eta_l) - 1]\, P_l(\cos\theta).} \tag{1.4.25}$$

It also follows from (1.4.3) and (1.4.24) that Ψ is given by

$$\Psi = \sum_{l=0}^{\infty} (2l+1) i^l \exp(i\eta_l) \frac{F_l(r)}{kr} P_l(\cos\theta). \tag{1.4.26}$$

We have thus succeeded in obtaining a solution of (1.4.1) satisfying (1.2.11) provided the series (1.4.25) converges for all θ and provided the series (1.4.26) converges for all r and θ.

An alternative way of writing (1.4.25) is

$$f(\theta) = \frac{1}{k} \sum_{l=0}^{\infty} (2l+1) \exp(i\eta_l) \sin \eta_l \, P_l(\cos \theta). \tag{1.4.27}$$

As is to be expected when U possesses spherical symmetry, the scattering amplitude is independent of φ.

By (1.2.9) the differential cross-section is given by

$$\sigma(\theta) = \frac{1}{2k^2} \sum_{l,n=0}^{\infty} (2l+1)(2n+1) \exp\{i(\eta_l - \eta_n)\} \times$$

$$\times \sin \eta_l \sin \eta_n \, P_l(\cos \theta) \, P_n(\cos \theta). \tag{1.4.28}$$

If we apply (1.1.2) and use (1.3.7) we see that the total cross section is given by

$$\sigma = 2\pi \int_0^{\pi} \sigma(\theta) \sin \theta \, d\theta = \frac{4\pi}{k^2} \sum_{l=0}^{\infty} (2l+1) \sin^2 \eta_l. \tag{1.4.29}$$

An interesting relation between the total cross-section σ and the *forward scattering amplitude* $f(0)$ is obtained if we put $\theta = 0$ in (1.4.27), take the imaginary part, and then use (1.4.29); it is

$$\sigma = \frac{4\pi}{k} \operatorname{Im} f(0). \tag{1.4.30}$$

This important result is known as the *Optical Theorem* and is essentially a consequence of the conservation of particles, as is shown in Appendix A.

To illustrate the results of this section, we give two examples.

EXAMPLE 1.2. *Quantum scattering by a hard sphere.* In Example 1.1 of Section 1.1 we considered classical scattering by a hard sphere. In the quantum mechanical analogue of this problem the potential will be given by

$$U = \frac{2\mu V}{\hbar^2} = \begin{cases} \infty, & r \leqq a, \\ 0, & r > a. \end{cases} \tag{1.4.31}$$

We must solve the radial equation (1.4.6) subject to the boundary condition (1.4.10). This will give us η_l, and then (1.4.28) and (1.4.29) enable us to obtain the differential and total cross-sections. For $r > a$, F_l satisfies (1.4.13), and it is shown in books on quantum mechanics that F_l also satisfies the boundary condition

$$F_l(a+) = 0 \tag{1.4.32}$$

where $F_l(a+)$ is the limit of $F_l(r)$ as $r \to a$ through values greater than a. We saw in Section 1.3 that the general solution of (1.4.3) is a linear combination of the solutions $krj_l(kr)$ and $krn_l(kr)$, and taking $A \cos \alpha$, $A \sin \alpha$ as our coefficients, F_l must have the form

$$F_l(r) = Akr[\cos \alpha\, j_l(kr) + \sin \alpha\, n_l(kr)]. \tag{1.4.33}$$

On use of (1.3.28) and (1.3.29), (1.4.33) becomes, as $r \to \infty$,

$$F_l(r) \underset{r\to\infty}{\sim} A \sin (kr - \tfrac{1}{2}l\pi - \alpha). \tag{1.4.34}$$

Comparison of this with (1.4.10) shows that $A = 1$, $\alpha = -\eta_l$, and so by (1.4.33),

$$F_l(r) = kr[\cos \eta_l\, j_l(kr) - \sin \eta_l\, n_l(kr)]. \tag{1.4.35}$$

Just as in the general case η_l is determined by the regularity of F_l at the origin, so in this case η_l is determined by the inner boundary condition, that is (1.4.32), and on taking the limit $r \to a+$ in (1.4.35) we get

$$\tan \eta_l = \frac{j_l(ka)}{n_l(ka)}. \tag{1.4.36}$$

This determines η_l, and hence $f(\theta)$ and σ from (1.4.27) and (1.4.29). For sufficiently low energies, when $k \to 0$, we can use (1.3.26) and (1.3.27) to approximate (1.4.36) by

$$\tan \eta_l \simeq -\frac{(ka)^{2l+1}}{(2l+1)\,[1.1.3.5.\cdots.(2l-1)]^2} \tag{1.4.37}$$

so that, if $ka \ll 1$, $\tan \eta_l \simeq \eta_l \simeq \sin \eta_l$. This allows us to replace (1.4.27) by

$$f(\theta) = -\frac{1}{k}\sum_{l=0}^{\infty} \frac{(ka)^{2l+1}}{[1.1.3.5.\cdots.(2l-1)]^2} \exp(i\eta_l)\, P_l(\cos\theta), \tag{1.4.38}$$

which is rapidly converging. Similarly (1.4.29) and (1.4.37) give us the total cross-section, viz.

$$\sigma \simeq \frac{4\pi}{k^2}\sum_{l=0}^{\infty} \frac{(ka)^{4l+2}}{(2l+1)\,[1.1.3.5.\cdots.(2l-1)]^4}, \tag{1.4.39}$$

again rapidly converging.

These results illustrate the general fact that for low energies only the first few terms of the partial wave expansion are required. If we consider electron scattering ($\mu = m$), and use atomic units (Vol. 1, p. 32: units of action, mass and length are $\hbar$, m, a, where a_0 is the Bohr radius), then $\hbar k = \mu v$ implies $k = v$. Taking $a \sim 1$, the condition $ka \ll 1$ becomes $v \ll 1$, that is $E = \frac{1}{2}\mu v^2 \ll \frac{1}{2}\mu$. The atomic unit of energy is 27·07 eV (electron-volts), hence for electron scattering we would expect the method to work well for energies of no more than a few eV. For protons, $\mu \sim 2000$ atomic units; hence the appropriate energy range is less than a few keV (1 keV = 1000 eV). In the limit as $k \to 0$, (1.4.39) gives us

$$\sigma = 4\pi a^2, \tag{1.4.40}$$

which is 4 times the classical cross-section obtained in Example 1.1. Since the condition $ka \ll 1$ is equivalent to saying that the wavelength $2\pi/k$ of the incident particles is very much larger than the radius a of the sphere, this result can be interpreted as due to diffraction effects.

Example 1.3. *Scattering by a square well potential.* In the case of a square well, the potential V is given by

$$V = \begin{cases} V_0, & r \leqq a, \\ 0, & r > a. \end{cases} \tag{1.4.41}$$

We see from (1.4.6) that, if $U_0 = 2\mu V_0/\hbar^2$ and $k^2 > U_0$, F_l must satisfy the differential equations

$$\left[\frac{d^2}{dr^2} - \frac{l(l+1)}{r^2} + k^2\right] F_l(r) = 0, \quad (r > a), \tag{1.4.42}$$

$$\left[\frac{d^2}{dr^2} - \frac{l(l+1)}{r^2} + \alpha^2\right] F_l(r) = 0, \quad (r < a), \tag{1.4.43}$$

where $\alpha = (k^2 - U_0)^{\frac{1}{2}}$. As in Example 1.2, (1.4.42) must have the solution

$$F_l(r) = kr[\cos \eta_l \, j_l(kr) - \sin \eta_l \, n_l(kr)], \quad (r > a), \tag{1.4.44}$$

where η_l is to be determined. Now F_l is to vanish at the origin and so (1.4.43), which is (1.4.42) with k replaced by α, has the solution

$$F_l(r) = A_l \, \alpha r \, j_l(\alpha r), \quad (r < a), \tag{1.4.45}$$

where A_l is a constant. Since it is required that a wave function and its first order derivative must be everywhere continuous, F_l and its derivative must be continuous. It saves labour if we note that this implies the continuity of

$$\left\{\frac{1}{r} F_l(r)\right\} \Big/ \left\{\frac{d}{dr}\frac{1}{r} F_l(r)\right\},$$

and so when $r = a$, (1.4.44) and (1.4.45) give

$$\frac{\cos \eta_l \, j_l(ka) - \sin \eta_l \, n_l(ka)}{k[\cos \eta_l \, j_l'(ka) - \sin \eta_l \, n_l'(ka)]} = \frac{j_l(\alpha a)}{\alpha j_l'(\alpha a)}. \tag{1.4.46}$$

We can solve this equation for η_l, and easily obtain

$$\eta_l = \tan^{-1}\left\{\frac{\alpha j_l(ka) \, j_l'(\alpha a) - k j_l(\alpha a) \, j_l'(ka)}{\alpha n_l(ka) \, j_l'(\alpha a) - k n_l'(ka) \, j_l(\alpha a)}\right\}. \tag{1.4.47}$$

Since we know η_l, we can calculate $f(\theta)$ and σ from (1.4.27) and (1.4.29).

1.5. The variational methods of Hulthén and Kohn

Variational methods play a very important part in the practical solution of many problems arising in physics and engineering; for example, the Ritz method for the solution of partial differential equations subject to certain boundary conditions (Elsgolc, 1961) and, of course, the variational technique for determining the bound states and discrete energies of atomic or molecular systems (Volume 2, Chapter 1). They have become particularly useful in recent years with the advent of modern high-speed computers. It is therefore not surprising that attempts have been made to develop such methods in the field of quantum scattering theory. In this section we shall consider one such method, which is particularly useful in the field of low-energy scattering when the method of partial waves discussed in Section 1.4 can be used.

Let us first recall the principal features of the most common variational method used for the solution of the bound state problem (Volume 2, Chapter 1). We first define a *functional* $E(\psi)$, which assigns to each *function* ψ of Hilbert space a *number* $E(\psi)$, according to the formula

$$E(\psi) = \frac{\langle\psi|\mathsf{H}|\psi\rangle}{\langle\psi|\psi\rangle}. \tag{1.5.1}$$

$E(\psi)$ is just the expectation value of the energy of the system when its wave function is ψ. If we replace ψ by a neighbouring function $\psi+\delta\psi$, we obtain a new value $E(\psi+\delta\psi)$ of the energy. Since $\delta\psi$ is small, we can calculate the difference $E(\psi+\delta\psi)-E(\psi)$ to first order in $\delta\psi$. This first-order difference is called the "variation" $\delta E(\psi)$ of the functional $E(\psi)$ corresponding to the change from ψ to $\psi+\delta\psi$. Now from (1.5.1) we have

$$E(\psi+\delta\psi)-E(\psi) = \frac{\langle\psi+\delta\psi|\mathsf{H}|\psi+\delta\psi\rangle}{\langle\psi+\delta\psi|\psi+\delta\psi\rangle} - \frac{\langle\psi|\mathsf{H}|\psi\rangle}{\langle\psi|\psi\rangle}$$

$$\simeq \frac{\langle\psi|\mathsf{H}|\psi\rangle+\langle\delta\psi|\mathsf{H}|\psi\rangle+\langle\psi|\mathsf{H}|\delta\psi\rangle}{\langle\psi|\psi\rangle+\langle\delta\psi|\psi\rangle+\langle\psi|\delta\psi\rangle} - \frac{\langle\psi|\mathsf{H}|\psi\rangle}{\langle\psi|\psi\rangle}$$

$$\simeq \frac{\langle\psi|\mathsf{H}|\psi\rangle+\langle\delta\psi|\mathsf{H}|\psi\rangle+\langle\psi|\mathsf{H}|\delta\psi\rangle}{\langle\psi|\psi\rangle}\left[1-\frac{\langle\delta\psi|\psi\rangle}{\langle\psi|\psi\rangle}-\frac{\langle\psi|\delta\psi\rangle}{\langle\psi|\psi\rangle}\right]$$

$$-\frac{\langle\psi|\mathsf{H}|\psi\rangle}{\langle\psi|\psi\rangle} \quad \text{(by the binomial theorem)}$$

$$\simeq \frac{\langle\delta\psi|\mathsf{H}|\psi\rangle+\langle\psi|\mathsf{H}|\delta\psi\rangle}{\langle\psi|\psi\rangle} - \frac{\langle\psi|\mathsf{H}|\psi\rangle\,[\langle\delta\psi|\psi\rangle+\langle\psi|\delta\psi\rangle]}{\langle\psi|\psi\rangle^2} \tag{1.5.2}$$

and since this is to first order in $\delta\psi$ and H is Hermitian

$$\delta E(\psi) = \frac{\langle\delta\psi|\mathsf{H}|\psi\rangle+\langle\mathsf{H}\psi|\delta\psi\rangle}{\langle\psi|\psi\rangle} - \frac{\langle\psi|\mathsf{H}|\psi\rangle\,[\langle\delta\psi|\psi\rangle+\langle\psi|\delta\psi\rangle]}{\langle\psi|\psi\rangle^2}. \tag{1.5.3}$$

We see immediately from (1.5.3) that if $\mathsf{H}\psi = E\psi$, $\delta E(\psi) = 0$, so that the bound states of H cause the variation of the energy functional to vanish. It now becomes possible to calculate bound states by considering trial functions ψ_t, which depend upon a finite set of parameters $\lambda_1, \lambda_2 \ldots \lambda_\nu$, and the condition for it to have a stationary value is

$$\partial E(\psi_t)/\partial\lambda_j = 0, \quad j = 1, 2 \ldots \nu. \tag{1.5.4}$$

Equation (1.5.4) represents a set of ν simultaneous equations which may be solved for $\lambda_1 \ldots \lambda_\nu$ to give an approximate wave function, while the corresponding energy is then given by $E(\psi_t)$, the value of the functional.

Let us now consider how a similar approach may be used to find approximate phase shifts for the scattering problem. We shall first consider s-wave scattering, when $l = 0$, for the generalization is very straightforward. The radial equation (1.4.6) now becomes

$$\left[\frac{d^2}{dr^2}+k^2-U(r)\right]F_0(r) = 0, \tag{1.5.5}$$

which has to be solved subject to the boundary condition (1.4.8) as $r \to \infty$ and regularity of the wave function at the origin, viz.

$$F_0(r) \underset{r\to\infty}{\sim} A_0 \sin(kr+\eta_0), \quad F_0(0) = 0. \tag{1.5.6}$$

We shall not, just yet, specify A_0; that is, we shall not consider F_0 to be normalized.

Consider now the functional

$$I(F) = \int_0^\infty F(r)\left[\frac{d^2}{dr^2}+k^2-U(r)\right]F(r)\,dr \tag{1.5.7}$$

where F satisfies similar boundary conditions to (1.5.6), viz.

$$F(r) \underset{r\to\infty}{\sim} A\sin(kr+\eta), \quad F(0) = 0. \tag{1.5.8}$$

The asymptotic condition (1.5.8) at infinity ensures that the integrand of (1.5.7) becomes small as $r \to \infty$. Provided it becomes sufficiently small (and this will always be the case with suitably chosen F), the integral will converge, and so the functional $I(F)$ will be defined.

Let us vary the function F to some neighbouring function $F+\delta F = \tilde{F}$ satisfying the boundary conditions

$$\tilde{F} = F+\delta F \underset{r\to\infty}{\sim} (A+\delta A)\sin(kr+\eta+\delta\eta), \quad \tilde{F}(0) = 0. \tag{1.5.9}$$

The variation $\delta I(F)$ is the value of $I(F+\delta F)-I(F)$ to first order, and so from (1.5.7) is equal to

$$\delta I(F) = \int_0^\infty \delta F\left[\frac{d^2}{dr^2}+k^2-U\right]F\,dr+\int_0^\infty F\left[\frac{d^2}{dr^2}+k^2-U\right]\delta F\,dr. \tag{1.5.10}$$

If we note that $(d/dr)\delta F = (d/dr)[\tilde{F}-F] = \tilde{F}'-F' = \delta F'$, an integration twice by parts gives

$$\delta I(F) = 2\int_0^\infty \delta F\left[\frac{d^2}{dr^2}+k^2-U\right]F\,dr+\Big[F\delta F'-F'\delta F\Big]_{r=0}^{r=\infty}. \tag{1.5.11}$$

Since $F(0) = \delta F(0) = 0$, use of (1.5.8) and (1.5.9) gives, to first order,

$$\begin{aligned}
\Big[F\delta F'-F'\delta F\Big]_{r=0}^{r=\infty}
&= \Big[F(F'+\delta F')-F'(F+\delta F)\Big]_{r=0}^{r=\infty}\\
&= \Big[F\tilde{F}'-F'\tilde{F}\Big]_{r=0}^{r=\infty}\\
&= \lim_{r\to\infty}\{A(A+\delta A)k\,[\sin(kr+\eta)\cos(kr+\eta+\delta\eta)\\
&\qquad -\cos(kr+\eta)\sin(kr+\eta+\delta\eta)]\}\\
&= -A(A+\delta A)k\sin\delta\eta \simeq -A^2k\delta\eta.
\end{aligned} \tag{1.5.12}$$

If we insert this into (1.5.11) we see that

$$\delta I(F) = 2\int_0^\infty \delta F\left[\frac{d^2}{dr^2}+k^2-U\right]F\,dr-A^2k\delta\eta, \tag{1.5.13}$$

and if $F = F_0$, so that it satisfies (1.5.5) and (1.5.6), we have

$$\delta I(F_0) = -A^2k\delta\eta. \tag{1.5.14}$$

We now restrict the class of functions over which we vary by imposition of the condition $A = \sec\eta$. Thus (1.5.8) becomes

$$F(r) \underset{r\to\infty}{\sim} \sin kr+\tan\eta\cos kr, \quad F(0) = 0, \tag{1.5.15}$$

while (1.5.14) shows that

$$\begin{aligned}\delta[I(F_0)+k \tan \eta] &= \delta I(F_0)+k\delta(\tan \eta) \\ &= \delta I(F_0)+k \sec^2 \eta \; \delta\eta \\ &= \delta I(F_0)+kA^2 \; \delta\eta \\ &= 0. \end{aligned} \tag{1.5.16}$$

If we define a new functional $J(F)$ by

$$J(F) = I(F)+k \tan \eta, \tag{1.5.17}$$

we see that (1.5.16) implies

$$\delta J(F_0) = 0. \tag{1.5.18}$$

Since F_0 satisfies (1.5.5), (1.5.7) shows that $I(F_0) = 0$, and so (1.5.17) gives

$$J(F_0) = k \tan \eta_0. \tag{1.5.19}$$

Hence the exact wave function F_0 makes the functional $J(F)$ stationary, and the phase shift is then given in terms of the stationary value of the functional by (1.5.19).

To apply the method in practice, we select a suitable trial wave function $F_t(\alpha_1, \alpha_2, \ldots, \alpha_\nu, \tan \eta)$ which depends parametrically upon the $\nu+1$ parameters $\alpha_1, \alpha_2, \ldots, \alpha_\nu, \tan \eta$, and also satisfies the boundary conditions (1.5.15). For example, in the case of electron–hydrogen scattering (the atom being treated as a point source of potential), Massey and Moiseiwitch (1951) have used a function essentially similar to the following

$$F_t(r) = \sin kr+[\tan \eta+\lambda_1 \exp(-r/a)]\,[1-\exp(-r/a)]\cos kr. \tag{1.5.20}$$

In this case $\nu = 1$, a being a given constant, and F_t clearly satisfies (1.5.15). When we substitute for F_t in (1.5.7), we can carry out the differentiation and integration to obtain I as a function of $\lambda_1, \ldots, \lambda_\nu$, $\tan \eta$. It therefore follows from (1.5.17) that $J(F)$ is also just a function of these $\nu+1$ variables. The stationary value is then given by the equations

$$\partial J/\partial \lambda_j = 0, \quad (j = 1, 2, \ldots, \nu); \quad \partial J/\partial(\tan \eta) = 0. \tag{1.5.21}$$

The approximate value η_0 (approx.) is then calculated from (1.5.19) with F_0 replaced by the function F_t (best) whose parameters satisfy (1.5.21); that is,

$$J[F_t \text{ (best)}] = k \tan [\eta_0 \text{ (approx.)}]. \tag{1.5.22}$$

The method just described is that of Kohn. Normally the trial function is linear in the parameters $\lambda_1, \ldots, \lambda_\nu$, $\tan \eta$, as, for example, is the case with the trial function given by (1.5.20). The expression (1.5.7) then shows that $I(F)$ is quadratic in the variable parameters, and so from (1.5.17) the function $J[F_t(\lambda_1, \ldots, \lambda_\nu, \tan \eta)]$ must also be quadratic. The equations (1.5.21) thus become $\nu+1$ linear equations for $\nu+1$ unknowns. The expression (1.5.22) for the phase shift is to be preferred to taking the value of η that occurs in F_t (best) for the following reason: if $F = F_0+\delta F$ is any function which is close to the exact solution F_0, $J(F_0+\delta F)-J(F_0)$ must be second order in δF [by (1.5.18)]. Hence by (1.5.19), $J(F_0+\delta F)-k \tan \eta_0$ is second order in δF; that is, $J(F_0+\delta F) = k \tan \eta_0$ with an error of second order. On the other hand, $\tan \eta - \tan \eta_0 = \delta(\tan \eta)$, and so is first order in δF.

There is an alternative approach due to Hulthén. If F_t were, in fact, the exact wave function F_0, it would satisfy

$$I(F_t) = 0, \tag{1.5.23}$$

for as we have already remarked in obtaining (1.5.19), $I(F_0) = 0$. Since the parameters $\lambda_1, \ldots, \lambda_\nu$, $\tan \eta$ already satisfy the $\nu+1$ equations (1.5.21), they would then satisfy $\nu+2$ equations. Now in general the class of functions of the form $F_t(\lambda_1, \ldots, \lambda_\nu, \tan \eta)$ does not contain F_0, and so in general the conditions (1.5.21) and (1.5.23) will give $\nu+2$ equations for $\nu+1$ unknowns, and these cannot all be satisfied. Kohn's method is to solve the equations (1.5.21) for the parameters. Hulthén's method consists in solving the $\nu+1$ equations

$$\partial J(F_t)/\partial \lambda_j = 0, \quad (j = 1, \ldots, \nu); \quad I(F_t) = 0. \tag{1.5.24}$$

Since $I(F_t) = 0$, it follows from (1.5.17) that $J(F_t) = k \tan \eta$, so that (1.5.19) is automatically satisfied by the phase of the best trial function. The method is, however, usually more complicated. For if the trial functions F_t are linear in the parameters, the equation $I(F_t) = 0$ will be quadratic. Equation (1.5.24) thus consists of $\nu+1$ simultaneous equations for the $\nu+1$ parameters, of which ν are linear but the remaining one is quadratic. They are therefore more difficult to solve than in the Kohn case, and there is an ambiguity of sign. For this reason Hulthén's method is used less frequently than that of Kohn.

The modifications to be made when $l \neq 0$ are very simple. The argument kr in the asymptotic formulae for F must be replaced by $kr - \frac{1}{2}l\pi$, while $U(r)$ must be replaced throughout by $U(r) - l(l+1)/r^2$. The reader who is interested in further details of the method is referred elsewhere (Demkov, 1963).

REFERENCES

CHURCHILL, R. V. (1963) *Fourier Series and Boundary Value Problems*, McGraw-Hill.

DEMKOV, YU. N. (1963) *Variational Principles in the Theory of Collisions*, Pergamon Press.

ELSGOLC, L. E. (1961) *Calculus of Variations*, Pergamon Press.

MANDL, F. (1957) *Quantum Mechanics*, Butterworth's Scientific Publications.

MASSEY, H. S. W. and MOISEIWITCH, B. L. (1951) *Proc. Roy. Soc.*, A, **205**, 483.

SNEDDON, I. N. (1961) *Special Functions of Mathematical Physics and Chemistry*, Oliver & Boyd.

WATSON, G. N. (1958) *A Treatise on the Theory of Bessel Functions*, Cambridge University Press.

CHAPTER 2

HIGH-ENERGY SCATTERING AND COULOMB SCATTERING

2.1. Green's function for a free particle

In Chapter 1 we formulated the quantum mechanical problem of the scattering of a particle by a fixed centre of force, and discussed its approximate solution by the method of partial waves and the associated variational techniques of Hulthén and Kohn. It became evident that these methods only work well when the energy of the incident particle is fairly small, and so in this chapter we shall seek a method which will enable us to perform reliable calculations at high energies. We shall also consider the separate problem of the scattering of a particle by Coulomb potential, when the formulation of Chapter 1 runs into difficulties.

The approximation appropriate for high energies is that due to Born, and this approximation is derived by first transforming the problem into one in integral equations. We do this by means of a "Green's function", and in this section we shall obtain the so-called free particle Green's function.

In the absence of an interaction the wave equation for a particle is (1.4.11), which involves the operator ∇^2+k^2. We now consider the related equation

$$(\nabla_{\mathbf{r}}^2+k^2)\, g(\mathbf{r},\, \mathbf{s}) = \delta(\mathbf{r}-\mathbf{s}) \tag{2.1.1}$$

where $\nabla_{\mathbf{r}}^2$ is the Laplacian operator ∇^2 relative to the coordinates (x, y, z) of $\mathbf{r}$, $\delta(\mathbf{r}-\mathbf{s})$ is the Dirac δ-function (see, for example, Volume 1, p. 112), and g is some as yet unspecified function of $\mathbf{r}$ and $\mathbf{s}$. There are many solutions of (2.1.1), for if $g_1(\mathbf{r}, \mathbf{s})$ is a solution and $g_2(\mathbf{r}, \mathbf{s})$ is any function which satisfies

$$(\nabla_{\mathbf{r}}^2+k^2)\, g_2(\mathbf{r},\, \mathbf{s}) = 0 \text{ everywhere,} \tag{2.1.2}$$

then g_1+g_2 is also a solution of (2.1.1).

It simplifies matters if for the time being we take a new origin at the fixed point $\mathbf{s}$, so that (2.1.1) becomes

$$(\nabla_{\mathbf{r}}^2+k^2)\, g(\mathbf{r},\, \mathbf{O}) = \delta(\mathbf{r}). \tag{2.1.3}$$

Now $g(\mathbf{r}, \mathbf{O})$ will satisfy (2.1.3) if and only if it satisfies the two equations

$$(\nabla_{\mathbf{r}}^2 + k^2)\, g(\mathbf{r}, \mathbf{O}) = 0, \quad \mathbf{r} \neq \mathbf{O}, \tag{2.1.4}$$

$$\int (\nabla_{\mathbf{r}}^2 + k^2)\, g(\mathbf{r}, \mathbf{O})\, d\mathbf{r} = 1, \tag{2.1.5}$$

where the integration in (2.1.5) is over all space.

A solution satisfying both conditions is

$$g(\mathbf{r}, \mathbf{O}) = -\frac{\exp(ikr)}{4\pi r}. \tag{2.1.6}$$

First, by using (1.3.11) and (1.3.12), we see that (2.1.4) is satisfied. Furthermore, the integrand in (2.1.5) vanishes outside a sphere V of radius ε and centre the origin, however small ε is, and so

$$\begin{aligned} &\int (\nabla_{\mathbf{r}}^2 + k^2)\, g(\mathbf{r}, \mathbf{O})\, d\mathbf{r} \\ &\quad = \lim_{\varepsilon \to 0} \int_V (\nabla_{\mathbf{r}}^2 + k^2)\, g(\mathbf{r}, \mathbf{O})\, d\mathbf{r} \\ &\quad = \lim_{\varepsilon \to 0} \int_V \nabla_{\mathbf{r}}^2 g(\mathbf{r}, \mathbf{O})\, d\mathbf{r}. \end{aligned}$$

If we formally apply the divergence theorem we obtain

$$\int (\nabla_{\mathbf{r}}^2 + k^2)\, g(\mathbf{r}, \mathbf{O})\, d\mathbf{r} = \lim_{\varepsilon \to 0} \int_S \nabla_{\mathbf{r}}\, g(\mathbf{r}, \mathbf{O}) \cdot d\mathbf{S} \tag{2.1.7}$$

where S is the surface of V. Now if $d\mathbf{S}$ subtends a solid angle $d\omega$ at the origin, we have, using (2.1.6),

$$\begin{aligned} \nabla_{\mathbf{r}}\, g(\mathbf{r}, \mathbf{O}) . d\mathbf{S} &= \left(\frac{d}{dr}\, g(\mathbf{r}, \mathbf{O}) \right)_{r=\varepsilon} \varepsilon^2\, d\omega \\ &= \left\{ \frac{\exp(ik\varepsilon)}{4\pi\varepsilon^2} - \frac{ik \exp(ik\varepsilon)}{4\pi\varepsilon} \right\} \varepsilon^2\, d\omega \end{aligned}$$

and so (2.1.7) gives

$$\begin{aligned} &\int (\nabla_{\mathbf{r}}^2 + k^2)\, g(\mathbf{r}, \mathbf{O})\, d\mathbf{r} \\ &\quad = \lim_{\varepsilon \to 0} \int \left\{ \frac{\exp(ik\varepsilon)}{4\pi} - ik\varepsilon\, \frac{\exp(ik\varepsilon)}{4\pi} \right\} d\omega \\ &\quad = \lim_{\varepsilon \to 0} \{\exp(ik\varepsilon) - ik\varepsilon \exp(ik\varepsilon)\} \\ &\quad = 1 \end{aligned}$$

which verifies (2.1.5), and hence establishes (2.1.3). Equation (2.1.1) now follows by returning to the old origin, when $\mathbf{r}$ is replaced by

$\mathbf{r}-\mathbf{s}$, so that (2.1.6) gives

$$g(\mathbf{r}, \mathbf{s}) = -\frac{\exp\{ik|\mathbf{r}-\mathbf{s}|\}}{4\pi|\mathbf{r}-\mathbf{s}|}. \tag{2.1.8}$$

The above arguments are given to make (2.1.8) plausible—they do not constitute a rigorous proof. A mathematically rigorous argument yields the same results.

The function g is known as *a Green's function for the operator* ∇^2+k^2, or as *a free particle Green's function*; because of certain special properties, it is often referred to as *the* free particle Green's function. The properties we note are, first, that it depends parametrically on the energy through k—this can be denoted, if necessary, by notations such as $g_k(\mathbf{r}, \mathbf{s})$, $g(\mathbf{r}, \mathbf{s}|E)$, or just $g(E)$; second, it is *symmetric*, that is

$$g(\mathbf{r}, \mathbf{s}) = g(\mathbf{s}, \mathbf{r}), \tag{2.1.9}$$

and so

$$\nabla_{\mathbf{s}}{}^2 g(\mathbf{r}, \mathbf{s}) = \nabla_{\mathbf{s}}{}^2 g(\mathbf{s}, \mathbf{r}) = \delta(\mathbf{s}-\mathbf{r})$$

or

$$\nabla_{\mathbf{s}}{}^2 g(\mathbf{r}, \mathbf{s}) = \delta(\mathbf{r}-\mathbf{s}). \tag{2.1.10}$$

Finally, we note that

$$\begin{aligned} |\mathbf{r}-\mathbf{s}| &= (r^2-2\mathbf{r}\cdot\mathbf{s}+s^2)^{\frac{1}{2}} \\ &= r\left(1-\frac{2\hat{\mathbf{r}}\cdot\mathbf{s}}{r}+\frac{s^2}{r^2}\right)^{\frac{1}{2}} \quad (\text{where } \hat{\mathbf{r}} = \mathbf{r}/r) \\ &= r\left(1-\frac{\hat{\mathbf{r}}\cdot\mathbf{s}}{r}+0\left(\frac{1}{r^2}\right)\right) \end{aligned}$$

where $0(1/r^N)$ means terms tending to zero as $r \to \infty$ at least as fast as $1/r^N$.

$$\therefore\ |\mathbf{r}-\mathbf{s}| = r-\hat{\mathbf{r}}\cdot\hat{\mathbf{s}}+0\left(\frac{1}{r}\right)$$

and so for large r, (2.1.8) gives

$$g(\mathbf{r}, \mathbf{s}) \underset{r\to\infty}{\sim} -\frac{\exp\{ikr-ik\hat{\mathbf{r}}\cdot\hat{\mathbf{s}}\}}{4\pi r}. \tag{2.1.11}$$

Equation (2.1.11) shows that g behaves at infinity like an outgoing wave. As we shall see, it is this property which gives it such importance.

2.2. The integral equation approach

In Section 1.2 we saw that, in order to solve the scattering problem, we must obtain a solution of the wave equation

$$(\nabla_{\mathbf{r}}^2+k^2)\,\psi(\mathbf{r}) = U\psi(\mathbf{r}) \tag{2.2.1}$$

subject to the boundary condition

$$\Psi(\mathbf{r}) \underset{r\to\infty}{\sim} \exp(ikz)+f(\theta,\varphi)\,\frac{\exp(ikr)}{r}. \tag{2.2.2}$$

We then attacked this problem in Section 1.4 by expanding Ψ in terms of partial waves, and found that this approach worked well for low energies. In this section we transform the problem into one in integral equations. What we shall show is that any solution $\Psi(\mathbf{r})$ of the "integral equation"

$$\Psi(\mathbf{r}) = \exp(ikz)+\int g(\mathbf{r},\mathbf{s})\,U(\mathbf{s})\,\Psi(\mathbf{s})\,d\mathbf{s}, \tag{2.2.3}$$

where as usual the integral is over all $\mathbf{s}$, is also a solution of (2.2.1) satisfying the boundary condition (2.2.2). The scattering problem is to find a solution of (2.2.1) and (2.2.2) and we will have achieved this if we can find a solution of (2.2.3).

In order to prove that (2.2.3) does indeed imply (2.2.1) and (2.2.2) we first note that if $g(\mathbf{r},\mathbf{s})$ is given by (2.1.8), and we formally differentiate under the integral sign and use (2.1.1), then

$$\begin{aligned}(\nabla_{\mathbf{r}}^2+k^2)&\int g(\mathbf{r},\mathbf{s})\,U(\mathbf{s})\,\Psi(\mathbf{s})\,d\mathbf{s}\\ &= \int(\nabla_{\mathbf{r}}^2+k^2)\,g(\mathbf{r},\mathbf{s})\,U(\mathbf{s})\,\Psi(\mathbf{s})\,d\mathbf{s}\\ &= \int\delta(\mathbf{r}-\mathbf{s})\,U(\mathbf{s})\,\Psi(\mathbf{s})\,d\mathbf{s}\\ &= U(\mathbf{r})\,\Psi(\mathbf{r}).\end{aligned} \tag{2.2.4}$$

If we operate on (2.2.3) with $\nabla_{\mathbf{r}}^2+k^2$ and make use of (2.2.4) and the fact that $(\nabla_{\mathbf{r}}^2+k^2)\exp(ikz) = 0$, we obtain (2.2.1). That (2.2.2) is also satisfied is seen if we take r large in (2.2.3) and use (2.1.11), for then

$$\Psi(\mathbf{r}) \underset{r\to\infty}{\sim} \exp(ikz)-\frac{\exp(ikr)}{4\pi r}\int\exp(-ik\hat{\mathbf{r}}\cdot\mathbf{s})\,U(\mathbf{s})\Psi(\mathbf{s})\,d\mathbf{s}. \tag{2.2.5}$$

Hence (2.2.2) is satisfied, and

$$f(\theta, \varphi) = f(\hat{\mathbf{r}}) = -\frac{1}{4\pi} \int \exp(-ik\hat{\mathbf{r}}\cdot\mathbf{s})\, U(\mathbf{s})\, \Psi(\mathbf{s})\, d\mathbf{s}. \qquad (2.2.6)$$

Summing up, we have the result:

If a wave function $\Psi(\mathbf{r})$ satisfies the integral equation (2.2.3), then it satisfies the wave equation (2.2.1) and the outgoing wave boundary condition (2.2.2).	(2.2.7)

It should be emphasized that the above arguments are given in order to make (2.2.7) plausible, and are in no way a proof.

Since $U = 2\mu V/\hbar^2$, (2.2.6) can be rewritten

$$f(\theta, \varphi) = f(\hat{\mathbf{r}}) = -\frac{\mu}{2\pi\hbar^2} \int \exp(-ik\hat{\mathbf{r}}\cdot\mathbf{s})\, V(\mathbf{s})\, \Psi(\mathbf{s})\, d\mathbf{s}. \qquad (2.2.8)$$

The initial wave vector $\mathbf{k}$ has magnitude k and lies in the z-direction. The final wave number is also k but the vector lies in the direction of the scattered particles; that is, in the direction of the outgoing unit vector $\hat{\mathbf{r}}$. The final wave vector can thus be denoted by $\mathbf{l} = k\hat{\mathbf{r}}$, where $l = k$ is required by energy conservation. The scattering amplitude f may now be written $f(\mathbf{k} \to \mathbf{l})$, and from (2.2.8)

$$f(\mathbf{k} \to \mathbf{l}) = -\frac{\mu}{2\pi\hbar^2} \int \exp(-i\mathbf{l}\cdot\mathbf{s})\, V(\mathbf{s})\, \Psi(\mathbf{s})\, d\mathbf{s},$$

or replacing the variable of integration $\mathbf{s}$ by $\mathbf{r}$

$$f(\mathbf{k} \to \mathbf{l}) = -\frac{\mu}{2\pi\hbar^2} \int \exp(-i\mathbf{l}\cdot\mathbf{r})\, V(\mathbf{r})\, \Psi(\mathbf{r})\, d\mathbf{r}. \qquad (2.2.9)$$

We shall denote the initial state $\exp(ikz) = \exp(i\mathbf{k}\cdot\mathbf{r})$ by Φ_{k}, and the final state $\exp(i\mathbf{l}\cdot\mathbf{r})$ by Φ_{l}. Equation (2.2.9) can then be written

$$\boxed{f(\Phi_{\mathrm{k}} \to \Phi_{\mathrm{l}}) = -\frac{\mu}{2\pi\hbar^2} \langle \Phi_{\mathrm{l}} | \mathsf{V} | \Psi \rangle} \qquad (2.2.10)$$

where $\langle \varphi | \mathsf{A} | \psi \rangle$ denotes the scalar product of two state vectors φ and $\mathsf{A}\psi$, A being an operator, and Ψ denotes the *scattering state* whose wave function is $\Psi(\mathbf{r})$. For notation and terminology, the reader is referred to Volume 1, Chapter 5.

2.3. The Born approximation

We shall now use the integral equation formulation of Section 2.2 to obtain an approximation valid at high energies. Since $U = 2\mu V/\hbar^2$ and $g(\mathbf{r}, \mathbf{s})$ is given by (2.1.8), the integral equation (2.2.3) can be written

$$\Psi(\mathbf{r}) = \exp(ikz) - \frac{\mu}{2\pi\hbar^2}\int \frac{\exp(ik|\mathbf{r}-\mathbf{s}|)}{|\mathbf{r}-\mathbf{s}|} V(\mathbf{s})\,\Psi(\mathbf{s})d\mathbf{s}. \tag{2.3.1}$$

If we put $\mathbf{R} = \mathbf{r}-\mathbf{s}$, so that $\mathbf{s} = \mathbf{r}-\mathbf{R}$, and take spherical polar coordinates with arbitrary polar axis, we find that

$$\begin{aligned}\int \frac{\exp(ik|\mathbf{r}-\mathbf{s}|)}{|r-\mathbf{s}|} V(\mathbf{s})\,\Psi(\mathbf{s})\,d\mathbf{s} &= \int \frac{\exp(ikR)}{R} V(\mathbf{r}-\mathbf{R})\,\Psi(\mathbf{r}-\mathbf{R})\,d\mathbf{R}\\ &= \int d\hat{\mathbf{R}} \int_0^\infty R^2\,dR\,\frac{\exp(ikR)}{R} V(\mathbf{r}-\mathbf{R})\,\Psi(\mathbf{r}-\mathbf{R}),\end{aligned} \tag{2.3.2}$$

where $d\hat{\mathbf{R}} = \sin\theta_{\mathbf{R}}\,d\theta_{\mathbf{R}}\,d\varphi_{\mathbf{R}}$ denotes the angular part of the integration over $\mathbf{R}$. As $k\to\infty$, the sine and cosine terms of the exponential in the integral over R will oscillate more frequently, so that eventually they will oscillate rapidly compared with the variation of the rest of the integrand (Fig. 2.1). It is therefore reasonable to suppose that, as $k\to\infty$, the integral over R tends to zero, and so by (2.3.1) $\Psi(\mathbf{r})\to\exp(ikz)$. Thus for high energies (k large),

$$\Psi(\mathbf{r}) \underset{k\to\infty}{\sim} \exp(ikz) = \exp(i\mathbf{k}\cdot\mathbf{r}) = \Phi_{\mathbf{k}}(\mathbf{r}) \tag{2.3.3}$$

and so from (2.2.9) the expression for the scattering amplitude becomes

$$f(\mathbf{k}\to\mathbf{l}) = -\frac{\mu}{2\pi\hbar^2}\int \exp(-i\mathbf{l}\cdot\mathbf{r})\,V(\mathbf{r})\exp(i\mathbf{k}\cdot\mathbf{r})\,d\mathbf{r}. \tag{2.3.4}$$

Equation (2.3.4) is known as the *Born approximation* for the scattering amplitude, and since we have seen that this becomes valid as $k\to\infty$, this will be a high-energy approximation—it is therefore complementary to the method of partial waves. Empirically it is found that the approximation gives good agreement with experiment for bombarding atoms with energies approaching 1 meV (1 million eV) and for electrons with energies of a few hundred eV or more.

The *momentum transfer* (the momentum lost by the incident particle) is $\hbar\mathbf{k}-\hbar\mathbf{l}$; denoting this by $\hbar\mathbf{q}$, (2.3.4) can be rewritten

$$f(\mathbf{q}) = -\frac{\mu}{2\pi\hbar^2}\int \exp(i\mathbf{q}\cdot\mathbf{r})\, V(\mathbf{r})\, d\mathbf{r}. \tag{2.3.5}$$

We see that the scattering amplitude is just proportional to the Fourier transform of the potential.

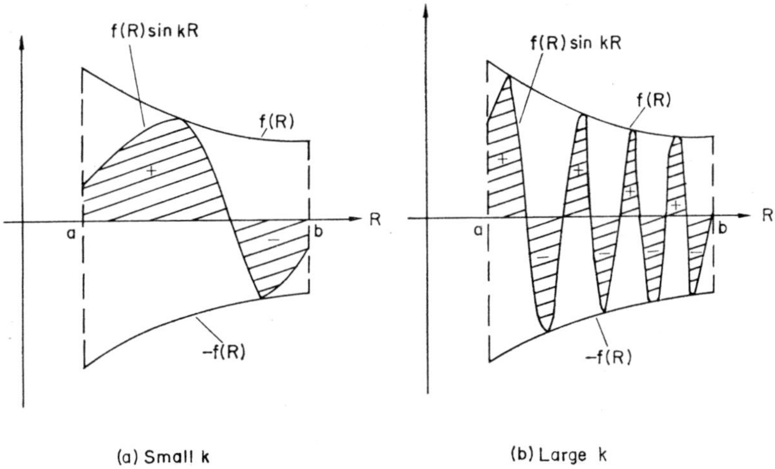

FIG. 2.1. $\int_a^b f(R) \sin kR\, dR$ is the shaded area. In (b) the areas of opposite sign tend to cancel, indicating that $\int_a^b f(R) \sin kR\, dR \to 0$ as $k \to \infty$.

If we remember that Φ_k and Φ_l denote the initial and final wave functions $\exp(i\mathbf{k}\cdot\mathbf{r})$ and $\exp(i\mathbf{l}\cdot\mathbf{r})$ respectively, (2.3.4) can be written

$$f(\Phi_\mathrm{k} \to \Phi_\mathrm{l}) = -\frac{\mu}{2\pi\hbar^2}\langle \Phi_\mathrm{l}|\mathrm{V}|\Phi_\mathrm{k}\rangle. \tag{2.3.6}$$

Comparison of (2.3.6) with the exact result (2.2.10) shows that the Born approximation is obtained by replacing the exact scattering state Ψ by the initial state Φ_k, a result which we shall see generalizes to the scattering of complex particles.

In many problems, the potential has spherical symmetry, so that V is a function of r only. Let us calculate the scattering amplitude in this case. To evaluate the integral in (2.3.5), choose $\mathbf{q}$ as polar axis of spherical polar coordinates $\mathbf{r} = (r, \xi, \eta)$, so that

$$f(\mathbf{q}) = -\frac{\mu}{2\pi\hbar^2}\int_0^\infty r^2\,dr \int_0^\pi \sin\xi\,d\xi \int_0^{2\pi} d\eta\, \exp(iqr\cos\xi)\, V(r)$$

$$= -\frac{\mu}{\hbar^2}\int_0^\infty r^2\,dr \int_{-1}^{+1} dt\, \exp(iqrt)\, V(r), \quad (t = \cos\xi).$$

If we carry out the integration over t, we find that

$$f(\mathbf{q}) = f(q) = -\frac{2\mu}{\hbar^2}\int_0^\infty \frac{\sin qr}{q}\, V(r)\, r\,dr. \tag{2.3.7}$$

We see from (2.3.7) that in the case of spherically symmetric problems the scattering amplitude depends only on the magnitude of the momentum transfer. Now $\mathbf{q} = \mathbf{k} - \mathbf{l}$, so

$$\begin{aligned} q^2 &= (\mathbf{k}-\mathbf{l})^2 \\ &= k^2 + l^2 - 2\mathbf{k}\cdot\mathbf{l} \\ &= k^2 + l^2 - 2kl\cos\theta, \end{aligned} \tag{2.3.8}$$

where θ is the angle between $\mathbf{k}$ and $\mathbf{l}$; in other words, the angle of scattering. Hence f is independent of the coordinate φ of $\mathbf{l}$ (see Fig. 2.2).

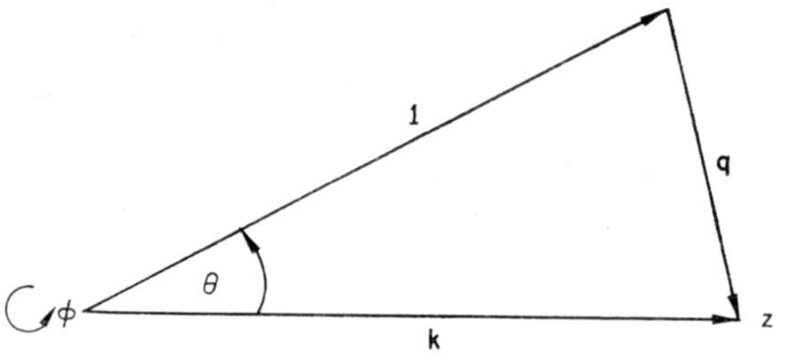

FIG. 2.2. Relation between initial and final wave numbers **k** and **l**, and momentum transfer $\hbar\mathbf{q}$.

Since $k = l$, (2.3.8) gives

$$q = 2k\sin\tfrac{1}{2}\theta, \tag{2.3.9}$$

and

$$q\,dq = k^2\sin\theta\,d\theta. \tag{2.3.10}$$

Equation (2.3.9) shows that q varies from 0 to $2k$ as θ varies from 0 to π.

Now from (1.1.2) and (1.2.9) we have

$$\boxed{\sigma = \int_0^{\pi} \sin\theta \, d\theta \int_0^{2\pi} d\varphi |f(\theta, \varphi)|^2.} \tag{2.3.11}$$

This is true quite generally, whether the Born approximation is valid or not, or whether V is spherically symmetric or not. In our present case of a spherically symmetric potential (2.3.11) becomes, on use of (2.3.10),

$$\sigma = \frac{2\pi}{k^2} \int_0^{2k} |f(q)|^2 \, q \, dq. \tag{2.3.12}$$

EXAMPLE 2.1. *Scattering by a screened Coulomb potential.* If an electron is scattered by, for example, a heavy atom, the electron cloud of the atom will "screen" the field of the nucleus. A possible potential for such scattering might be

$$V(r) = -\frac{Ze^2 \exp(-\alpha r)}{\kappa_0 r}, \quad (\alpha > 0), \tag{2.3.13}$$

where (using SI units) $\kappa_0 = 4\pi\varepsilon_0$ and ε_0 is the permittivity of free space. If mixed Gaussian units are employed this factor is absent. α is the screening constant, and Ze the nuclear charge. We can obtain the scattering amplitude $f(q)$ by substituting for V from (2.3.13) into (2.3.7). The reader may easily evaluate the resulting integral to find that

$$f(q) = +\left(\frac{2\mu Ze^2}{\kappa_0 \hbar^2}\right) \frac{1}{\alpha^2 + q^2}. \tag{2.3.14}$$

The differential cross-section is therefore

$$\sigma(\theta) = |f(q)|^2 = \left(\frac{2\mu Ze^2}{\kappa_0 \hbar^2}\right)^2 \frac{1}{(\alpha^2 + q^2)^2} \tag{2.3.15}$$

and the reader may obtain the total cross-section from (2.3.12). With the aid of (2.3.9), (2.3.15) may be written

$$\sigma(\theta) = \left(\frac{2\mu Ze^2}{\kappa_0 \hbar^2}\right)^2 \frac{1}{(\alpha^2 + 4k^2 \sin^2 \frac{1}{2}\theta)^2}. \tag{2.3.16}$$

In the limit as $\alpha \to 0$, this becomes

$$\sigma(\theta) = \left(\frac{\mu Ze^2}{2\kappa_0 \hbar^2 k^2}\right)^2 \operatorname{cosec}^4 \tfrac{1}{2}\theta,$$

or

$$\sigma(\theta) = \left(\frac{Ze^2}{2\kappa_0 \mu v^2}\right)^2 \operatorname{cosec}^4 \tfrac{1}{2}\theta, \tag{2.3.17}$$

which is the classical scattering formula first obtained by the famous British physicist Lord Rutherford. The absence of $\hbar$ from (2.3.17) is the reason for the coincidence of the classical and quantum mechanical formulae. The quantum mechanical result has here been derived in the limit of high energies, when the Born approximation presumably becomes exact. Since the classical formula is exact, one would expect (2.3.17) to be valid quantum mechanically at all energies. This is in fact the case, as a treatment of Coulomb scattering shows (see Section 2.5).

2.4. The Born series

The work of the last section showed us how to obtain an approximate solution of the integral equation (2.3.1) for high energies, when the potential has the effect of a small perturbation. This suggests that we might improve our results by solving (2.3.1) by an iteration procedure, and so obtain cross-sections which should be valid for lower energies.

It will be more convenient if we define a new Green's function $G_0(\mathbf{r}, \mathbf{s})$ by

$$G_0(\mathbf{r}, \mathbf{s}) = \frac{2\mu}{\hbar^2} g(\mathbf{r}, \mathbf{s}) = -\left(\frac{\mu}{2\pi\hbar^2}\right) \frac{\exp(ik|\mathbf{r}-\mathbf{s}|)}{|\mathbf{r}-\mathbf{s}|} \tag{2.4.1}$$

[cf. (2.1.8)] so that the integral equation (2.3.1) can be written

$$\Psi(\mathbf{r}) = \exp(ikz) + \int G_0(\mathbf{r}, \mathbf{s}) \, V(\mathbf{s}) \, \Psi(\mathbf{s}) \, d\mathbf{s}. \tag{2.4.2}$$

Let us recall (Volume 1, Chapter 3) that an *operator* A maps some function f into some other function $\mathsf{A}f$, so that to define A we have to specify the function $\mathsf{A}f$ into which any function f is mapped. To specify a function we must give its value at an arbitrary point $\mathbf{r}$. We will now define the *integral operator* G_0 by saying that, given the function f, the value $\mathsf{G}_0 f(\mathbf{r})$ of the function $\mathsf{G}_0 f$ at the arbitrary point $\mathbf{r}$ is given in terms of the values of f by

$$\mathsf{G}_0 f(\mathbf{r}) = \int G_0(\mathbf{r}, \mathbf{s}) f(\mathbf{s}) \, d\mathbf{s}. \tag{2.4.3}$$

The potential V is also an operator, $\mathsf{V}f$ being defined for all f according to

$$\mathsf{V}f(\mathbf{r}) = V(\mathbf{r}) f(\mathbf{r}); \tag{2.4.4}$$

it is a so-called "local" potential. Hence the operator $\mathsf{G}_0\mathsf{V}$ is given by

$$\mathsf{G}_0\mathsf{V}f(\mathbf{r}) = \int G_0(\mathbf{r}, \mathbf{s}) \, V(\mathbf{s}) f(\mathbf{s}) \, d\mathbf{s}. \tag{2.4.5}$$

Now if A is any operator such that, if f and g are any two functions, and λ, μ are any two complex numbers, then

$$\mathsf{A}(\lambda f + \mu g) = \lambda \mathsf{A} f + \mu \mathsf{A} g, \tag{2.4.6}$$

we say that A is a linear operator. The reader may easily verify that V, G_0, and G_0V are linear operators. Also, the product AB of two operators A and B is defined by

$$(\mathsf{AB})f = \mathsf{A}(\mathsf{B}f) \tag{2.4.7}$$

for any function f.

Since $\Phi_{\mathbf{k}}(\mathbf{r}) = \exp(ikz)$, the above definitions show that (2.4.2) can be written

$$\boxed{\Psi = \Phi_{\mathbf{k}} + \mathsf{G}_0\mathsf{V}\Psi.} \tag{2.4.8}$$

This equation can be solved formally by iteration; thus, since G_0V is linear,

$$\begin{aligned}\Psi &= \Phi_{\mathbf{k}} + \mathsf{G}_0\mathsf{V}(\Phi_{\mathbf{k}} + \mathsf{G}_0\mathsf{V}\Psi)\\ &= \Phi_{\mathbf{k}} + \mathsf{G}_0\mathsf{V}\Phi_{\mathbf{k}} + \mathsf{G}_0\mathsf{V}\mathsf{G}_0\mathsf{V}(\Phi_{\mathbf{k}} + \mathsf{G}_0\mathsf{V}\Psi);\end{aligned}$$

and repeated substitution gives

$$\Psi = \Phi_{\mathbf{k}} + \mathsf{G}_0\mathsf{V}\Phi_{\mathbf{k}} + \mathsf{G}_0\mathsf{V}\mathsf{G}_0\mathsf{V}\Phi_{\mathbf{k}} + \ldots, \tag{2.4.9}$$

a result known as the *Born series* for the wave function Ψ.

The solution (2.4.9) of (2.4.8) may be obtained in an alternative way. We remember that the sum or difference of two operators A and B is defined by

$$(\mathsf{A} \pm \mathsf{B})f = \mathsf{A}f \pm \mathsf{B}f \tag{2.4.10}$$

for any function f, so that (2.4.8) may be written

$$(\mathsf{I} - \mathsf{G}_0\mathsf{V})\Psi = \Phi_{\mathbf{k}} \tag{2.4.11}$$

where I is the *unit operator*, so that $\mathsf{I}f = f$. Equation (2.4.11) can be solved formally by

$$\Psi = (\mathsf{I} - \mathsf{G}_0\mathsf{V})^{-1}\,\Phi_{\mathbf{k}}, \tag{2.4.12}$$

and formally expanding the right-hand side by the binomial theorem, this gives

$$\Psi = (\mathsf{I} + \mathsf{G}_0\mathsf{V} + \mathsf{G}_0\mathsf{V}\mathsf{G}_0\mathsf{V} + \ldots)\,\Phi_{\mathbf{k}} \tag{2.4.13}$$

which is equivalent to (2.4.9).

To obtain the scattering amplitude we substitute for Ψ from (2.4.13) into (2.2.10) to obtain

$$f(\Phi_{\mathbf{k}} \to \Phi_{\mathbf{l}}) = -\frac{\mu}{2\pi\hbar^2}\langle\Phi_{\mathbf{l}}|\mathsf{V}|\Phi_{\mathbf{k}}\rangle - \frac{\mu}{2\pi\hbar^2}\langle\Phi_{\mathbf{l}}|\mathsf{V}\mathsf{G}_0\mathsf{V}|\Phi_{\mathbf{k}}\rangle - \ldots \tag{2.4.14}$$

which is an infinite series. Taking the first term of this series only gives us the Born approximation, which is therefore sometimes known as the *first Born approximation.* Taking the first two terms in (2.4.14) gives the *second Born approximation* for the scattering amplitude. Since, however,

$$\langle \Phi_l | V G_0 V | \Phi_k \rangle$$
$$= \int d\mathbf{r}\ \Phi_l(\mathbf{r})\ V(\mathbf{r})\ G_0 V \Phi_k(\mathbf{r})$$
$$= \int d\mathbf{r}\ \Phi_l(\mathbf{r})\ V(\mathbf{r}) \int d\mathbf{s}\ G_0(\mathbf{r}, \mathbf{s})\ V(\mathbf{s})\ \Phi_k(\mathbf{s}), \qquad (2.4.15)$$

this involves an integration over six variables. Often the results are not worth the labour involved.

The above manipulations are formal. Such formal manipulations are much used by physicists in deriving new results and approximations. Their justification, however, raises formidable mathematical difficulties; for a discussion, reference may be made to Mott and Massey (1965).

2.5. Coulomb scattering‡

In Section 2.3 we deduced the validity of Rutherford's scattering formula (2.3.17) in the high energy limit, and stated that it is in fact valid at all energies. In order to establish this formula we need to evaluate the scattering amplitude for the Coulomb potential. Unfortunately the treatment of earlier sections breaks down for potentials which tend to zero as $r \to \infty$ as slowly as, or more slowly than, $1/r$. In this section we examine the special treatment required for the Coulomb potential, deduce the Rutherford scattering formula, and discuss the reasons for the breakdown of the earlier treatments.

The potential for scattering of a charged particle of charge $Z_1 e$ by a fixed charge $Z_2 e$ placed at the origin is $Z_1 Z_2 e^2/\kappa_0 r$. The Schrödinger equation for the collision problem is then

$$\left[-\frac{\hbar^2}{2\mu}\nabla^2 + \frac{Z_1 Z_2 e^2}{\kappa_0 r}\right]\Psi(\mathbf{r}) = E\Psi(\mathbf{r}), \qquad (2.5.1)$$

which may be written

$$\left[\nabla^2 + k^2 - \frac{2\gamma k}{r}\right]\Psi(\mathbf{r}) = 0, \qquad (2.5.2)$$

where

$$\gamma = \mu Z_1 Z_2 e^2/\kappa_0\, \hbar^2 k = Z_1 Z_2 e^2/\kappa_0\, \hbar v \qquad (2.5.3)$$

We put

$$\Psi(\mathbf{r}) = \exp(ikz) f, \qquad (2.5.4)$$

‡ The reader who is not specifically interested in Coulomb scattering may omit this section, which is added primarily for the sake of completeness.

and substitute for Ψ in (2.5.2) to obtain

$$\nabla^2 f + 2ik\frac{\partial f}{\partial z} - \frac{2\gamma k}{r} f = 0. \tag{2.5.5}$$

This possesses a solution for f in the form $g(r-z)$, as we now show.

Since $\partial r/\partial x = x/r$, $\partial r/\partial y = y/r$, $\partial r/\partial z = z/r$, we obtain

$$\nabla^2 f = \frac{\partial}{\partial x}\left(g'\frac{\partial r}{\partial x}\right) + \frac{\partial}{\partial y}\left(g'\frac{\partial r}{\partial y}\right) + \frac{\partial}{\partial z}\left[g'\left(\frac{\partial r}{\partial z} - 1\right)\right]$$

$$= \left[\left(\frac{\partial r}{\partial x}\right)^2 + \left(\frac{\partial r}{\partial y}\right)^2 + \left(\frac{\partial r}{\partial z} - 1\right)^2\right] g'' + \left[\frac{3}{r} - \frac{x^2+y^2+z^2}{r^3}\right] g'$$

and hence

$$\nabla^2 f = 2g'' - 2g''\frac{z}{r} + \frac{2g}{r}. \tag{2.5.6}$$

It is also clear that

$$\frac{\partial f}{\partial z} = \left(\frac{z}{r} - 1\right) g'. \tag{2.5.7}$$

If we put (2.5.6) and (2.5.7) into (2.5.5) we get

$$(r-z)g'' + [1 - ik(r-z)]g' - \gamma k g = 0. \tag{2.5.8}$$

Let us now introduce a new variable u, related to the polar angle θ:

$$u = ik(r-z) = 2ikr \sin^2 \tfrac{1}{2}\theta. \tag{2.5.9}$$

In terms of u, noting that $g(r-z) = f$, and dividing (2.5.8) by ik, we find that

$$u\frac{d^2 f}{du^2} + (1-u)\frac{df}{du} + i\gamma f = 0. \tag{2.5.10}$$

Equation (2.5.10) is a special case of the confluent hypergeometric equation (see Section 1.3). Solutions regular at the origin must be proportional to the confluent hypergeometric function

$$_1F_1(-i\gamma, 1; u) = \sum_{n=0}^{\infty} \frac{(-i\gamma)_n}{(1)_n}\frac{u^n}{n!} \tag{2.5.11}$$

where the symbol $(\beta)_n$ is defined as

$$(\beta)_n = \beta(\beta+1)\ldots(\beta+n-1) = \frac{\Gamma(\beta+n)}{\Gamma(\beta)}, \tag{2.5.12}$$

Γ is the gamma function, and $(\beta)_0$ is defined as 1; for example, $(1)_n = n!$. That (2.5.11) does indeed represent a solution of (2.5.10) may be checked by solution in series; and it is obviously regular at the origin (where

$u = 0$). Equations (2.5.4) and (2.5.9) then show that the wave function $\Psi(\mathbf{r})$ is given by

$$\Psi(\mathbf{r}) = A \exp(ikz)\,{}_1F_1[-i\gamma, 1; ik(r-z)], \tag{2.5.13}$$

where A is a constant.

It may be shown (e.g. Lebedev, 1965) that

$$\begin{aligned} {}_1F_1(-i\gamma, 1; u) \underset{|u|\to\infty}{\sim} & \frac{1}{\Gamma(1+i\gamma)} \exp(\pm\pi\gamma)\, u^{i\gamma}\left[1+\gamma^2 u^{-1}+0\left(\frac{1}{u^2}\right)\right] \\ & + \frac{1}{\Gamma(-i\gamma)} \exp(u)\, u^{-(1+i\gamma)}\left[1+\frac{(1+i\gamma)^2}{u}+0\left(\frac{1}{u^2}\right)\right]. \end{aligned} \tag{2.5.14}$$

In this case $u = ik(r-z)$, hence $\mathrm{Im}(u) > 0$, and so the + sign must be chosen. Equations (2.5.13) and (2.5.14) show that

$$\begin{aligned} \Psi(\mathbf{r}) \underset{|r-z|\to\infty}{\sim} & \frac{A \exp(\pi\gamma)}{\Gamma(1+i\gamma)} \exp\{ikz+i\gamma \log[ik(r-z)]\}\left[1+\frac{\gamma^2}{ik(r-z)}+\ldots\right] \\ & + \frac{A}{\Gamma(-i\gamma)} \frac{\exp\{ikr-i\gamma\log[ik(r-z)]\}}{ik(r-z)}\left[1+\frac{(1+i\gamma)^2}{ik(r-z)}+\ldots\right]. \end{aligned}$$

Hence using (2.5.9)

$$\begin{aligned} \Psi(\mathbf{r}) \underset{|r-z|\to\infty}{\sim} & \frac{A \exp(\frac{1}{2}\pi\gamma)}{\Gamma(1+i\gamma)}\Bigg\{\exp[ikz+i\gamma\log(\overline{kr-z})]\left[1+\frac{\gamma^2}{ik(r-z)}+\ldots\right] \\ & + \frac{\Gamma(1+i\gamma)}{\Gamma(-i\gamma)} \frac{\exp[ikr-i\gamma\log(\overline{kr-z})]}{2ikr\sin^2\frac{1}{2}\theta} \times \\ & \times\left[1+\frac{(1+i\gamma)^2}{ik(r-z)}+\ldots\right]\Bigg\}. \end{aligned} \tag{2.5.15}$$

The first term in the curly brackets on the right-hand side of (2.5.15) represents a plane wave modified by the phase factor $i\gamma \log(\overline{kr-z})$, and on using (2.5.9) we see that the second term is asymptotically of the form $f_c(\theta) \exp[ikr-i\gamma\log(2kr)]/r$ where

$$f_c(\theta) = \frac{\Gamma(1+i\gamma)}{i\Gamma(-i\gamma)} \frac{\exp[-i\gamma\log(\sin^2\frac{1}{2}\theta)]}{2k\sin^2\frac{1}{2}\theta}. \tag{2.5.16}$$

The expression $\exp[ikr-i\gamma\log(2kr)]/r$ may be interpreted as an outgoing wave modified by the long-range nature of the Coulomb potential, and so $f_c(\theta)$ may be interpreted as the scattering amplitude. Since

$-i\gamma\Gamma(-i\gamma) = \Gamma(1-i\gamma) = \Gamma^*(1+i\gamma)$, (2.5.16) may be written

$$f_c(\theta) = -\frac{\gamma}{2k\sin^2\frac{1}{2}\theta}\exp\left[-i\gamma\log\left(\sin^2\tfrac{1}{2}\theta\right)+2i\sigma_0\right] \quad (2.5.17)$$

where $\sigma_0 = \arg\Gamma(1+i\gamma)$. The differential cross-section is given by

$$\sigma(\theta) = |f_c(\theta)|^2 = \left(\frac{\gamma}{2k}\right)^2 \operatorname{cosec}^4 \tfrac{1}{2}\theta$$

and hence, using (2.5.3),

$$\sigma(\theta) = \left(\frac{Z_1 Z_2 e^2}{2\mu v^2 \kappa_0}\right)^2 \operatorname{cosec}^4 \tfrac{1}{2}\theta. \quad (2.5.18)$$

This result is in agreement with Rutherford's scattering formula (2.3.17), to which it reduces when $Z_1 = -1$, $Z_2 = Z$. This confirms the remark made in Section 2.3 that agreement between classical and quantum mechanics is to be expected in this case.

It is characteristic of the Coulomb potential that the incident and scattered waves are distorted right out to infinity, as could have been anticipated by the following argument. The radial equation (1.4.6) may be written

$$\frac{d^2F_l}{dr^2}+[k^2-W(r)]F_l = 0, \quad (2.5.19)$$

where

$$W(r) = U(r)+\frac{l(l+1)}{r^2}. \quad (2.5.20)$$

Put

$$F_l = A\exp\left[\int_a^r f(r')\,dr'\right]\exp(\pm ikr). \quad (2.5.21)$$

Substitution for F_l from (2.5.21) into (2.5.19) gives

$$f'+f^2\pm 2ikf = W. \quad (2.5.22)$$

Now suppose that $W = 0(r^{-s})$ as $r\to\infty$ $(s>0)$, and suppose that if $f = 0(1/r^N)$ as $r\to\infty$ then $f' = 0(1/r^{N+1})$. Then (2.5.22) requires that $f = 0(1/r^s)$ as $r\to\infty$. Hence if $s>1$, (2.5.21) gives

$$F_l \underset{r\to\infty}{\sim} \text{const}\times\exp(\pm ikr), \quad (2.5.23)$$

and taking an appropriate combination of $\exp(ikr)$ and $\exp(-ikr)$ we obtain the asymptotic condition (1.4.10), from which the existence of a solution satisfying the boundary condition (1.2.11) was deduced in Section 1.4.

For Coulomb scattering,

$$W(r) \underset{r\to\infty}{\sim} -\frac{2\gamma k}{r}, \tag{2.5.24}$$

and so (2.5.22) gives

$$f \underset{r\to\infty}{\sim} \pm\frac{i\gamma}{r}. \tag{2.5.25}$$

Hence (2.5.21) gives

$$F_l(r) \underset{r\to\infty}{\sim} A \exp(\pm ikr \pm i\gamma \log r + \text{const}). \tag{2.5.26}$$

Equation (2.5.26) shows that the wave function asymptotically has the form of incoming and outgoing waves modified by the phase factor $\exp(\pm i\gamma \log r)$, and this may be compared with the asymptotic form (2.5.15).

A systematic and rigorous treatment of Coulomb scattering has still not been obtained, and there remain many problems concerning the treatment of collisions in which either initially there are two charged particles, or finally there are at least two charged particles.

REFERENCES

LEBEDEV, N. N. (1965) *Special Functions and their Applications*, Prentice-Hall.
MOTT, N. F. and MASSEY, H. S. W. (1965) *The Theory of Atomic Collisions*, Clarendon Press, Oxford.

CHAPTER 3

SCATTERING OF TWO STRUCTURELESS PARTICLES

3.1. Scattering of two distinguishable particles

In the first two chapters we considered the special case of scattering by a fixed centre of force. This is a good model of, for example, the scattering of an electron by a hydrogen atom, assuming that we know the force of interaction. In the case of proton–hydrogen scattering, however, the model breaks down; if the hydrogen atoms are the target particles, they will certainly be moving after the collision. In this section and the next we consider the problem of the scattering of two particles of comparable masses, assuming that we know the force (i.e. potential) between them. The remainder of the chapter will be taken up with the complications arising from the Pauli principle.

We shall label the colliding particles 1 and 2, and suppose them to have masses m_1 and m_2, velocities $\mathbf{v}_1$ and $\mathbf{v}_2$, momenta $\hbar\mathbf{k}_1 = m_1\mathbf{v}_1$ and $\hbar\mathbf{k}_2 = m\mathbf{v}_2$, and position vectors $\mathbf{r}_1$ and $\mathbf{r}_2$ respectively. We also suppose that the potential V is a function of $\mathbf{r}_1-\mathbf{r}_2$ only; then $V = V(\mathbf{r}_1-\mathbf{r}_2) = V(\mathbf{r})$, where $\mathbf{r} = \mathbf{r}_1-\mathbf{r}_2$ is the position vector of 1 relative to 2. In the case of classical scattering with a potential of this form, it leads to simplifications in the mathematics if we take new coordinates

$$\mathbf{R} = \frac{m_1\mathbf{r}_1+m_2\mathbf{r}_2}{m_1+m_2}, \quad \mathbf{r} = \mathbf{r}_1-\mathbf{r}_2, \tag{3.1.1}$$

so that $\mathbf{R}$ is the position vector of the centre of mass of the system, and we shall see that this is true also in the quantum mechanical case. Now the wave function of the unperturbed state in the laboratory system is

$$\Phi_G(\mathbf{r}_1, \mathbf{r}_2) = \exp(i\mathbf{k}_1\cdot\mathbf{r}_1)\exp(i\mathbf{k}_2\cdot\mathbf{r}_2) \tag{3.1.2}$$

where the suffix G indicates that this includes the motion of the centre of mass G. If we define $\mathbf{K}$ and $\mathbf{k}$ by

$$\mathbf{K} = \mathbf{k}_1+\mathbf{k}_2, \quad \mathbf{k} = \frac{m_2\mathbf{k}_1-m_1\mathbf{k}_2}{m_1+m_2} \tag{3.1.3}$$

so that $\hbar\mathbf{K}$ is the total momentum, (3.1.2) may be rewritten

$$\Phi_G(\mathbf{R}, \mathbf{r}) = \exp(i\mathbf{K}\cdot\mathbf{R})\exp(i\mathbf{k}\cdot\mathbf{r}). \tag{3.1.4}$$

The *reduced mass* μ of 1 and 2 is defined by

$$\mu = \frac{m_1 m_2}{m_1+m_2} \tag{3.1.5}$$

and we therefore see from (3.1.3) that $\hbar\mathbf{k} = \mu(\mathbf{v}_1-\mathbf{v}_2) = \mu\mathbf{v}$ say. Since $\mathbf{v} = \mathbf{v}_1-\mathbf{v}_2$ is the velocity of 1 relative to 2, $\hbar\mathbf{k}$ is the *relative momentum* of the particles, and $\mathbf{k}$ the *relative wave vector*.

The total Hamiltonian is

$$\mathsf{H}_G = -\frac{\hbar^2}{2m_1}\nabla_1{}^2-\frac{\hbar^2}{2m_2}\nabla_2{}^2+V(\mathbf{r}_1-\mathbf{r}_2) \tag{3.1.6}$$

where $\nabla_1{}^2$ is the Laplacian for $\mathbf{r}_1 = (x_1, y_1, z_1)$ so that

$$\nabla_1{}^2 = \frac{\partial^2}{\partial x_1{}^2}+\frac{\partial^2}{\partial y_1{}^2}+\frac{\partial^2}{\partial z_1{}^2} \tag{3.1.7}$$

and $\nabla_2{}^2$ is the Laplacian for $\mathbf{r}_2 = (x_2, y_2, z_2)$. By a simple transformation of coordinates (3.1.6) can be rewritten

$$\mathsf{H}_G = \mathsf{K}_G+H \tag{3.1.8}$$

where

$$\mathsf{K}_G = -\frac{\hbar^2}{2M}\nabla_{\mathbf{R}}{}^2 \tag{3.1.9}$$

is the kinetic energy operator associated with the motion of the centre of mass and

$$\mathsf{H} = -\frac{\hbar^2}{2\mu}\nabla_{\mathbf{r}}{}^2+V(\mathbf{r}) \tag{3.1.10}$$

is the Hamiltonian of the relative motion. $M = m_1+m_2$ is the total mass, $\nabla_{\mathbf{R}}{}^2$ is the Laplacian for $\mathbf{R}$, and $\nabla_{\mathbf{r}}{}^2$ is the Laplacian for $\mathbf{r}$. The unperturbed Hamiltonian H_{0G} is defined by

$$\mathsf{H}_{0G} = -\frac{\hbar^2}{2m_1}\nabla_1{}^2-\frac{\hbar^2}{2m_2}\nabla_2{}^2 = -\frac{\hbar^2}{2M}\nabla_{\mathbf{R}}{}^2-\frac{\hbar^2}{2\mu}\nabla_{\mathbf{r}}{}^2, \tag{3.1.11}$$

and so by (3.1.2) or (3.1.4)

$$\mathsf{H}_{0G}\,\Phi_G = \left(\frac{\hbar^2K^2}{2M}+\frac{\hbar^2k^2}{2\mu}\right)\Phi_G = \left(\frac{\hbar^2k_1{}^2}{2m_1}+\frac{\hbar^2k_2{}^2}{2m_2}\right)\Phi_G. \tag{3.1.12}$$

The expression multiplying Φ_G in (3.1.12) is the total energy E_G, say, and this therefore has the form

$$E_G = E_K + E \tag{3.1.13}$$

where $E_K \equiv \hbar^2 K^2/2M$ is the energy associated with the centre of mass, and $E = \hbar^2 k^2/(2\mu) = \frac{1}{2}\mu v^2$ is the *relative energy* of the particles.

The centre of mass motion is unperturbed by the presence of the potential $V(\mathbf{r})$ (as in the classical case) and so the total scattering state Ψ_G can be written

$$\Psi_G(\mathbf{R}, \mathbf{r}) = \exp(i\mathbf{K}\cdot\mathbf{R})\,\Psi(\mathbf{r}) \tag{3.1.14}$$

where

$$\mathsf{H}\Psi = E\Psi. \tag{3.1.15}$$

Ψ is the wave function for relative motion, and for large r it must have the form of a plane wave and an outgoing spherical wave describing the motion of the incident particle 1 relative to the target particle 2. Hence

$$\Psi(\mathbf{r}) \underset{r\to\infty}{\sim} \exp(i\mathbf{k}\cdot\mathbf{r}) + f(\hat{\mathbf{r}})\frac{\exp(ikr)}{r}. \tag{3.1.16}$$

We have thus reduced the scattering problem to that of Chapter 1 by separating out the centre of mass motion.

3.2. Relation between the centre of mass and laboratory systems

In the relative coordinate system the incident flux is $\hbar k/\mu = v$, and so we can define the differential cross-section in this system by

$$\boxed{\sigma(\theta, \varphi) = |f(\theta, \varphi)|^2.} \tag{3.2.1}$$

This is not the cross-section measured by the experimentalist. He in fact measures $\sigma_l(\theta, \varphi)$, the differential cross-section in the laboratory system where the target particles are initially at rest; since the flux in the laboratory system is still v, this will be $1/v$ times the number of particles appearing in unit solid angle per unit time in the direction specified by the angles θ_l, φ_l, say. The two systems are illustrated in Fig. 3.1, where we show particle 2 as the target and particle 1 as incident, and so in the laboratory system 1 initially has velocity v and 2 is at rest. We suppose that the direction θ_l, φ_l and element of solid angle $d\omega_l = \sin\theta_l\, d\theta_l\, d\varphi_l$ in the laboratory system corresponds to the direction θ, φ and element of solid angle $d\omega = \sin\theta\, d\theta\, d\varphi$ in the relative coordinate system. In order to obtain the relation between the two systems, we look at the centre of mass system which is obtained from the laboratory system by imposing on it a velocity $v_2 = m_1 v/(m_1+m_2)$ to the left. This brings the centre of mass to rest, and since the centre of mass is unaffected by the collision we obtain the situation shown in Fig. 3.1(b). The initial velocities in the centre of mass system are

$$m_2: \quad v_2 = \frac{m_1 v}{m_1+m_2} \text{ to the left,} \tag{3.2.2}$$

$$m_1: \quad v_1 = v - v_2 = \frac{m_2 v}{m_1+m_2} \text{ to the right.} \tag{3.2.3}$$

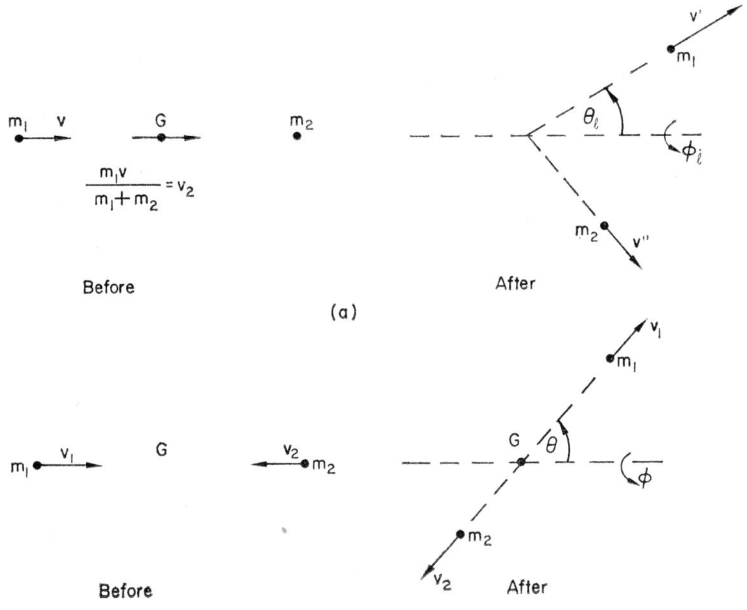

(a)

(b)

FIG. 3.1. Laboratory and centre of mass systems. (a) Laboratory system. (b) Centre of mass system.

Now $v_2/v_1 = m_1/m_2 = \gamma$, say; hence if v_1 and v_2 are changed by the collision, they either both increase or both decrease. We suppose that the collision is *elastic*; in other words, that energy is conserved. It therefore follows that v_1 and v_2 can neither both increase nor both decrease, and so they are unchanged by the collision, as shown in Fig. 3.1(b). Furthermore, the initial and final directions of particle 1 are the same in the centre of mass and relative coordinate systems, as also shown in Fig. 3.1(b).

We shall suppose as before that the flux of incident particles 1 is I; let us suppose further that there are N particles 2 in the target chamber. By the obvious extension of the definition of the term cross-section to this case we see that the number of particles appearing in the solid angle $d\omega$ per unit time is equal to $NI\sigma(\theta, \varphi)d\omega$, and this will be the same as the number $NI\sigma_l(\theta_l, \varphi_l)d\omega_l$ appearing per unit time in $d\omega_l$; hence

$$\sigma(\theta, \varphi)d\omega = \sigma_l(\theta_l, \varphi_l)d\omega_l. \tag{3.2.4}$$

In order to relate θ_l and φ_l with θ and φ we use the vector diagram of Fig. 3.2, from which we see that $\varphi_l = \varphi$ (scattering takes place in the same plane through Oz for both systems) and

$$v_1 \sin\theta = v' \sin\theta_l, \tag{3.2.5}$$

$$v_1 \cos\theta + v_2 = v' \cos\theta_l. \tag{3.2.6}$$

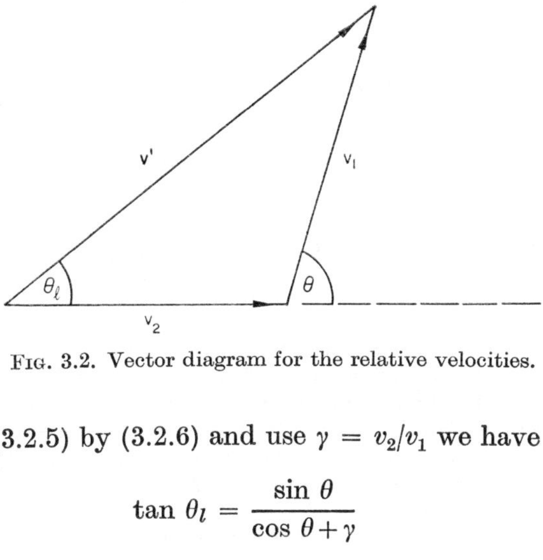

FIG. 3.2. Vector diagram for the relative velocities.

If we divide (3.2.5) by (3.2.6) and use $\gamma = v_2/v_1$ we have

$$\tan\theta_l = \frac{\sin\theta}{\cos\theta + \gamma} \tag{3.2.7}$$

and hence

$$\cos\theta_l = \frac{\cos\theta + \gamma}{(1 + 2\gamma\cos\theta + \gamma^2)^{\frac{1}{2}}}. \tag{3.2.8}$$

The sign in (3.2.8) follows by continuity, for in the limit $m_2 \to \infty$ so that $\gamma \to 0$ the laboratory and relative coordinate systems will coincide, and then certainly $\cos\theta = \cos\theta_l$. Since the sign cannot alter as γ is varied continuously from zero, it must remain positive. (When $\gamma = -\cos\theta$, the derivative of $\cos\theta$ with respect to $\cos\theta$ must not change sign.)

From (3.2.4) we have

$$\sigma_l(\theta_l, \varphi_l) \left|\sin \theta_l \, d\theta_l \, d\varphi_l\right| = \sigma(\theta, \varphi) \left|\sin \theta \, d\theta \, d\varphi\right|, \tag{3.2.9}$$

the modulus signs appearing since it is the magnitude of corresponding solid angles which enters into (3.2.4).

$$\therefore \ \sigma_l(\theta_l, \varphi_l) = \sigma(\theta, \varphi) \left|\frac{d(\cos \theta)}{d(\cos \theta_l)}\right|. \tag{3.2.10}$$

If we substitute for $\cos \theta_l$ from (3.2.8) into (3.2.10) we find that

$$\sigma_l(\theta_l, \varphi_l) = \frac{(1+2\gamma \cos \theta + \gamma^2)^{3/2}}{|1+\gamma \cos \theta|} \sigma(\theta, \varphi). \tag{3.2.11}$$

Equation (3.2.11) enables us to calculate the observed differential cross-section in terms of the differential cross-section calculated in the relative coordinate system. The total cross-section is just the number of particles scattered per unit time, and is therefore the same in both systems.

In particular, if the two particles have the same mass, $\gamma = 1$; then (3.2.7) gives $2\theta_l = \theta$, and so (3.2.11) becomes

$$\sigma_l(\theta_l, \varphi_l) = 4 \cos \theta_l \, \sigma(2\theta_l, \varphi_l). \tag{3.2.12}$$

If $m_2 = \infty$ so that the target particle 2 remains at rest, $\gamma = 0$, and we see from (3.2.7) and (3.2.11) that $\theta_l = \theta$, $\sigma_l(\theta_l \ \varphi_l) = \sigma(\theta, \varphi)$, as we should expect.

EXAMPLE 3.1. *Classical scattering of two hard spheres.* We suppose the two spheres to have the same radius "a" and same mass, and that the collision is perfectly elastic. With a little thought the reader will convince himself that the problem in the centre of mass system is the same as that for scattering of a point particle by a hard sphere of radius $2a$, and so we can apply (1.1.7) with a replaced by $2a$. It follows that $\sigma(\theta, \varphi) = a^2$, and so (3.2.12) gives

$$\sigma_l(\theta_l, \varphi_l) = 4a^2 \cos \theta_l \tag{3.2.13}$$

for the differential cross-section in the laboratory system. The total cross-section is

$$\sigma = \int_0^{\pi} \sin \theta \, d\theta \int_0^{2\pi} d\varphi \, \sigma(\theta, \varphi) = 4\pi a^2 \tag{3.2.14}$$

as is to be expected. In the laboratory system $\theta_l = \frac{1}{2}\theta$ and varies from 0 to $\frac{1}{2}\pi$, and hence

$$\sigma = \int_0^{\frac{1}{2}\pi} \sin \theta_l \, d\theta_l \int_0^{2\pi} d\varphi_l \, . \, 4a^2 \cos \theta_l = 4\pi a^2 \tag{3.2.15}$$

confirming that the total cross-section is the same in both systems in this case.

3.3. Scattering of two identical spinless particles

In this section we shall consider the complications arising when the two particles are identical. Before dealing with the quantum mechanical case, let us first consider the classical situation.

Referring to Fig. 3.1(b), we recall that, in the centre of mass system, the number of incident particles emerging in unit solid angle $d\omega$ per unit time per unit incident flux is $\sigma_d(\theta, \varphi)d\omega$, where we add the suffix to denote *direct scattering*; that is to say, we count only the *incident* particles emerging in the direction θ, φ. In addition to these, to each incident particle leaving in the direction $(\pi-\theta, \varphi+\pi)$ there corresponds a recoil target particle emerging in the direction (θ, φ), and so the number of recoil particles emerging in $d\omega$ per unit time is $\sigma_d(\pi-\theta, \varphi+\pi)d\omega$. If the particles are identical the detecting counter will in general not distinguish between incident and recoil particles, and the number of particles emerging in $d\omega$ per unit time per unit flux is $\sigma(\theta, \varphi)d\omega$ where $\sigma(\theta, \varphi)$ is the *observed* differential cross-section. Thus $\sigma(\theta, \varphi)$ is given by

$$\sigma(\theta, \varphi) = \sigma_d(\theta, \varphi)+\sigma_d(\pi-\theta, \varphi+\pi). \tag{3.3.1}$$

In most experimental arrangements the target particles are usually at rest, and since $\gamma = 1$ we can apply (3.2.12) to obtain

$$\sigma_l(\theta_l, \varphi_l) = 4 \cos \theta_l[\sigma_d(2\theta_l, \varphi_l)+\sigma_d(\pi-2\theta_l, \varphi_l+\pi)]. \tag{3.3.2}$$

EXAMPLE 3.2. *Classical scattering of two alpha particles.* The expression for $\sigma(\theta, \varphi)$ is now given by the Rutherford formula (2.5.18); if M is the mass of a proton (or neutron), the mass of an α-particle (helium nucleus) is $4M$, and so $\mu = 2M$, whilst $Z_1 = Z_2 = 2$. Thus

$$\sigma_d(\theta, \varphi) = \left(\frac{e^2}{Mv^2\kappa_0}\right)^2 \operatorname{cosec}^4 \tfrac{1}{2}\theta \tag{3.3.3}$$

and so (3.3.2) gives

$$\sigma(\theta_l, \varphi_l) = 4 \cos \theta_l \left(\frac{e^2}{Mv^2\kappa_0}\right)^2 [\operatorname{cosec}^4 \theta_l+\sec^4 \theta_l]. \tag{3.3.4}$$

We now consider the quantum mechanical case. The α-particles discussed above are examples of particles known as bosons.

It is a well-established fact (based on a wealth of experimental evidence) that the wave function describing two or more identical bosons must be symmetric in all its coordinates. Hence (3.1.2) is not strictly acceptable and instead we should use a function of the form

$$\Phi_G(\mathbf{r}_1, \mathbf{r}_2|\mathscr{S}) = \exp(i\mathbf{k}_1\cdot\mathbf{r}_1)\exp(i\mathbf{k}_2\cdot\mathbf{r}_2)+\exp(i\mathbf{k}_1\cdot\mathbf{r}_2)\exp(i\mathbf{k}_2\cdot\mathbf{r}_1) \tag{3.3.5}$$

which describes two beams of particles, one with momentum $\hbar\mathbf{k}_1$ and the other with momentum $\hbar\mathbf{k}_2$. By use of the change of coordinates (3.1.1) we can rewrite (3.3.5) as

$$\Phi_G(\mathbf{r}, \mathbf{R}|\mathscr{S}) = \exp(i\mathbf{K}\cdot\mathbf{R})\,[\exp(i\mathbf{k}\cdot\mathbf{r})+\exp(-i\mathbf{k}\cdot\mathbf{r})]. \qquad (3.3.6)$$

We note that the motion of the centre of mass is unaffected, as before, whilst if $\mathbf{\Psi}(\mathbf{r})$ is the solution of the Schrödinger equation

$$\left[-\frac{\hbar^2}{2\mu}\nabla_r^2+V(\mathbf{r})\right]\mathbf{\Psi}(\mathbf{r}) = E\mathbf{\Psi}(\mathbf{r}) \qquad (3.3.7)$$

satisfying the boundary condition

$$\mathbf{\Psi}(\mathbf{r}) \underset{r\to\infty}{\sim} \exp(i\mathbf{k}\cdot\mathbf{r})+f(\hat{\mathbf{r}})\frac{\exp(ikr)}{r} \qquad (3.3.8)$$

then the symmetrized wave function

$$\mathbf{\Psi}(\mathbf{r}|\mathscr{S}) = \mathbf{\Psi}(\mathbf{r})+\mathbf{\Psi}(-\mathbf{r}) \qquad (3.3.9)$$

also satisfies (3.2.7) (since $\nabla^2_{-\mathbf{r}} = \nabla^2_{\mathbf{r}}$ and $V(-\mathbf{r}) = V(\mathbf{r})$) and the boundary condition

$$\mathbf{\Psi}(\mathbf{r}|\mathscr{S}) \underset{r\to\infty}{\sim} \exp(i\mathbf{k}\cdot\mathbf{r})+\exp(-i\mathbf{k}\cdot\mathbf{r})+[f(\hat{\mathbf{r}})+f(-\hat{\mathbf{r}})]\,\frac{\exp(ikr)}{r}. \qquad (3.3.10)$$

The expression $f(\hat{\mathbf{r}})+f(-\hat{\mathbf{r}})$ is the amplitude for the scattering of one particle into the direction $\hat{\mathbf{r}}$ whilst the other recoils in the direction $-\hat{\mathbf{r}}$, and the differential cross-section will be given by

$$\begin{aligned}\sigma(\theta, \varphi) &= c|f(\hat{\mathbf{r}})+f(-\hat{\mathbf{r}})|^2\\ &= c|f(\theta, \varphi)+f(\pi-\theta, \varphi+\pi)|^2\\ &= c[|f(\theta, \varphi)|^2+|f(\pi-\theta, \varphi+\pi)|^2\\ &\qquad +2\mathrm{Re}\{f^*(\theta, \varphi)f(\pi-\theta, \varphi+\pi)\}]. \qquad (3.3.11)\end{aligned}$$

In (3.3.11) Re means real part, and we have introduced the positive constant c since it is not clear that $\mathbf{\Psi}(\mathscr{S})$ is correctly normalized. In fact $c = 1$, for if there is no back scattering in the centre of mass system, $f(\theta, \varphi)$ will vansh unless θ is acute, and so the cross term in (3.3.11) vanishes; hence

$$\sigma(\theta, \varphi) = c[|f(\theta, \varphi)|^2+|f(\pi-\theta, \varphi+\pi)|^2]. \qquad (3.3.12)$$

Equation (3.3.12) shows that $\sigma(\theta, \varphi) = c|f(\theta, \varphi)|^2$ when θ is acute, and $\sigma(\theta, \varphi) = c|f(\pi-\theta, \varphi+\pi)|^2$ when θ is obtuse, so that if θ is acute we *know* that the scattered particle is the incident particle, and if θ is obtuse we *know* that the scattered particle is the recoil particle. That is to say, we have the case of distinguishable particles, and so we expect the classical result (3.3.1); hence $c = 1$. Thus we obtain the result

$$\boxed{\text{for bosons: } \sigma(\theta, \varphi) = |f(\theta, \varphi)+f(\pi-\theta, \varphi+\pi)|^2.}\,^{\ddagger} \tag{3.3.13}$$

The differential cross-section in the laboratory system is again given by (3.2.12) and so from (3.3.13)

$$\sigma_l(\theta_l, \varphi_l) = 4 \cos \theta_l |f(2\theta_l, \varphi_l)+f(\pi-2\theta_l, \varphi_l+\pi)|^2. \tag{3.3.14}$$

EXAMPLE 3.3. *Quantum scattering of two α-particles.* The scattering amplitude is given by (2.5.17) with $\gamma = Z_1 Z_2 e^2/\kappa_0\hbar v = 4e^2/\kappa_0\hbar v$; since $\hbar k = \mu v = 2Mv$ we see from (3.3.14) that

$$\sigma_l(\theta_l, \varphi_l) = 4 \cos \theta_l \left(\frac{e^2}{\kappa_0 M v^2}\right)^2 \times$$
$$\times \left[\operatorname{cosec}^4 \theta_l + \sec^4 \theta_l + 2 \operatorname{cosec}^2 \theta_l \sec^2 \theta_l \cos\left\{\frac{4e^2}{\kappa_0 \hbar v} \log \tan^2 \theta_l\right\}\right]. \tag{3.3.15}$$

Equation (3.3.15) differs from the classical result (3.3.4) due to the presence of the interference term; when $\theta_l = 45°$, the quantum cross-section is twice the classical one. This gives a direct experimental test of the quantum theory, and the results of experiments on the scattering of α-particles confirm (3.3.15).

If the α-particles were screened, so that the potential had the form (2.3.13) with $Z = -4$, and we applied the Born approximation, we could use (2.3.14); since $q = 2k \sin \frac{1}{2}\theta$ we obtain

$$f(\theta, \varphi) = -\frac{16Me^2}{\kappa_0 \hbar^2} \frac{1}{(\alpha^2+4k^2 \sin^2 \frac{1}{2}\theta)}. \tag{3.3.16}$$

If we substitute for $f(\theta, \varphi)$ from (3.3.16) into (3.3.14) and let $\alpha \to 0$ we obtain

$$\sigma_l(\theta_l, \varphi_l) = 4 \cos \theta_l \left[\frac{e^2}{\kappa_0 M v^2}\right]^2 [\operatorname{cosec}^4 \theta_l + \sec^4 \theta_l + 2 \operatorname{cosec}^2 \theta_l \sec^2 \theta_l]. \tag{3.3.17}$$

Equation (3.3.17) differs both from the classical result (3.3.4) and the accurate quantum mechanical result (3.3.15), but has been obtained by application of the Born approximation. If we let $v \to \infty$, we see that (3.3.15) and (3.3.17) become similar, confirming that in this case the Born approximation becomes accurate in the limit of high energies.

‡ The above argument is not entirely satisfactory and this point will be taken up later when the wave packet approach is discussed.

At high energies the particles will only be slightly scattered. Hence $f(\theta, \varphi)$ will vanish unless θ is small and therefore acute, and for these values of θ it follows from (3.3.13) that $\sigma(\theta, \varphi) \simeq |f(\theta, \varphi)|^2$. This is usually true in the case of more complex collisions. At high energies exchange of particles is unlikely to take place, and so symmetrization of the overall wave function for the system is then usually unnecessary.

3.4. Scattering of two identical particles with spin

The last section dealt with the scattering of two identical spinless particles such as α-particles. We will now consider the modifications to be made when the particles have a fourth degree of freedom, namely spin.

For simplicity we shall confine our attention to spin $\frac{1}{2}$ particles (Volume 1, Section 4.9) such as electrons and protons, although the discussion is easily generalized. Let s denote the spin coordinate; then the *spin space* corresponding to one particle is a two-dimensional vector space. Suppose α, β are the spin functions defined by

$$\alpha(+\tfrac{1}{2}) = 1,\ \alpha(-\tfrac{1}{2}) = 0; \quad \beta(+\tfrac{1}{2}) = 0,\ \beta(-\tfrac{1}{2}) = 1; \tag{3.4.1}$$

then α, β form an orthonormal basis for the spin space of one particle. α is the wave function of a particle with spin $\frac{1}{2}\hbar$ in the direction Oz, and β is the wave function for a particle with spin $-\frac{1}{2}\hbar$ in the direction Oz.

The spin space of two particles is the "direct product" of the spin spaces of the separate particles. If we label the particles 1 and 2, and let s_1, s_2 be the spin coordinates of 1 and 2, respectively, then the set

$$\alpha(s_1)\,\alpha(s_2), \quad \alpha(s_1)\,\beta(s_2), \quad \beta(s_1)\,\alpha(s_2), \quad \beta(s_1)\,\beta(s_2), \tag{3.4.2}$$

forms an orthonormal set spanning the spin space for the two particles. We shall henceforth refer to this set with the abbreviated notation

$$\alpha(1)\,\alpha(2), \quad \alpha(1)\,\beta(2), \quad \beta(1)\,\alpha(2), \quad \beta(1)\,\beta(2). \tag{3.4.3}$$

We can select linear combinations of the members of (3.4.3) to form the orthonormal sets

$$\left\{\begin{array}{c} \alpha(1)\,\alpha(2), \\ \dfrac{1}{\sqrt{2}}\,[\alpha(1)\,\beta(2)+\alpha(2)\,\beta(1)], \\ \beta(1)\,\beta(2), \end{array}\right\} \tag{3.4.4}$$

$$\frac{1}{\sqrt{2}}\,[\alpha(1)\,\beta(2)-\alpha(2)\,\beta(1)]. \tag{3.4.5}$$

The functions (3.4.4) and (3.4.5) together form an orthonormal basis for the spin space of the two particles. The set (3.4.4) spans the *symmetric* subspace of the product space, whilst (3.4.5) spans the antisymmetric subspace.

In order to take account of the spatial coordinates, we must form the direct product of the four-dimensional spin space with the infinite dimensional space of wave functions $\Psi(\mathbf{r}_1, \mathbf{r}_2)$, which we shall abbreviate to $\Psi(1, 2)$. Spin $\frac{1}{2}$ particles are called fermions and, in contrast with the bosons (p. 51), their wave functions are antisymmetric in the particle coordinates (with spins included). Hence if $u_i(\mathbf{r})$ is a complete orthogonal set of spatial wave functions, we must use as our *physically allowed* basis functions the set (3.4.4) multiplied by $u_i(1)\,u_j(2)-u_i(2)\,u_j(1)$ and the set (3.4.5) multiplied by $u_i(1)\,u_j(2)+u_i(2)\,u_j(1)$. In particular, the set (3.4.4) must be multiplied by the scattering state $\Psi(\mathbf{r})-\Psi(-\mathbf{r})$ in the relative coordinate system, and (3.4.5) must be multiplied by $\Psi(\mathbf{r})+\Psi(-\mathbf{r})$. Corresponding to each of the initial states we have a scattering amplitude $f(\theta, \varphi)\pm f(\pi-\theta, \varphi+\pi)$, the minus sign being taken for each of the states (3.4.4) and the plus sign being taken for (3.4.5). Assuming that each of these states occurs with equal probability, the differential cross-section is given by

$$\sigma(\theta, \varphi) = \tfrac{1}{4}|f(\theta, \varphi)+f(\pi-\theta, \varphi+\pi)|^2+\tfrac{3}{4}|f(\theta, \varphi)-f(\pi-\theta, \varphi+\pi)|^2. \tag{3.4.6}$$

From (3.2.12) we see that in the laboratory system

$$\sigma_l(\theta_l, \varphi_l) = 4\cos\theta_l\,[|f(2\theta_l, \varphi_l)|^2+|f(\pi-2\theta_l, \varphi_l+\pi)|^2 - \operatorname{Re}\{f^*(2\theta_l, \varphi_l)\,f(\pi-2\theta_l, \varphi_l+\pi)\}]. \tag{3.4.7}$$

Example 3.4. *Scattering of two electrons.* The amplitude is given by (2.5.17) with $\gamma = e^2/\hbar v\kappa_0$ and if the mass of an electron is m, $\mu = \frac{1}{2}m$, $\gamma/2k = e^2/mv^2\kappa_0$, so that (3.4.7) gives

$$\sigma_l(\theta_l, \varphi_l) = 4\cos\theta_l\left(\frac{e^2}{mv^2\kappa_0}\right)^2 \times \left[\operatorname{cosec}^4\theta_l+\sec^4\theta_l-\operatorname{cosec}^2\theta_l\sec^2\theta_l\cos\left(\frac{e^2}{\hbar v\kappa_0}\log\tan^2\theta_l\right)\right] \tag{3.4.8}$$

which has been confirmed by cloud chamber experiments. The observations are made by observing the scattering of high energy electrons by atomic electrons, the binding and motion of the atomic electrons being negligible at high energies of 20 keV.

REFERENCE

Mandl, F. (1957) *Quantum Mechanics,* Butterworth's Scientific Publications.

CHAPTER 4

SCATTERING OF TWO COMPLEX PARTICLES

4.1. Direct collisions

We must now take up the problem of the collision of two particles whose internal structure cannot be ignored; for example, the collision of two atoms or molecules. Let us denote the colliding systems by A and B, and concentrate on what are called *direct collisions*. These are of the form

$$A+B \rightarrow A+B; \tag{4.1.1}$$

that is to say, the complex particles collide and then separate without any exchange of constituent particles. Direct collisions are of two types: *elastic* and *inelastic*. In elastic collisions the internal states of A and B are unchanged—only the relative velocity is altered, but in an inelastic collision the internal states of one or both particles are changed. We shall assume that energy is conserved, so that in the case of an elastic collision the magnitude of the relative velocity is constant, and only its direction is altered. If the collision is inelastic and the internal energy is increased as the result of the excitation of higher internal states of the particles, the relative speed v must be correspondingly reduced. On the other hand, if the overall internal energy is reduced the relative speed must increase. Since the combined centre of mass of A and B moves freely, the energy E_K associated with its motion is constant, and only the relative kinetic energy is affected by an inelastic collision.

Let us now consider how we describe the initial state of the system. Each complex particle A and B is described by wave functions χ_{Ar} and χ_{Bs} respectively, where r and s stand for the quantum numbers of their initial states. χ_{Ar} and χ_{Bs} are functions of the internal coordinates $\mathbf{x}_A$ and $\mathbf{x}_B$ of A and B respectively. If the internal coordinates $\mathbf{x}_A$ and $\mathbf{x}_B$ are combined and denoted by $\mathbf{x}$, say, the internal state of A and B may be described by the combined wave function χ_n where

$$\chi_n(\mathbf{x}) = \chi_{Ar}(\mathbf{x}_A)\, \chi_{Bs}(\mathbf{x}_B). \tag{4.1.2}$$

This is clearly more convenient than writing out the internal states separately. If h_A and h_B are the internal Hamiltonians of A and B, $\mathsf{h} \equiv \mathsf{h}_A+\mathsf{h}_B$ is the internal Hamiltonian of the system. Hence if E_r, E_s

are the internal energies of A and B respectively, so that $E_n \equiv E_r + E_s$ is the total internal energy, we clearly have

$$\mathsf{h}_A\, \chi_{Ar} = E_r\, \chi_{Ar}, \quad \mathsf{h}_B\, \chi_{Bs} = E_s\, \chi_{Bs}, \quad \mathsf{h}\, \chi_n = E_n\, \chi_n. \tag{4.1.3}$$

Initially the centre of mass of A moves with momentum $\hbar\mathbf{k} = \mu\mathbf{v}$ relative to the centre of mass of B, where $\mathbf{k}$ is the relative wave vector, μ the reduced mass, and $\mathbf{v}$ the relative velocity of the centres of mass. Such a motion is naturally described by the plane wave $\exp(i\mathbf{k}\cdot\mathbf{r})$, where $\mathbf{r}$ is now the position vector of the centre of mass of A relative to the centre of mass of B. The overall initial state for the relative motion is therefore described by the product of χ_n and a plane wave. We shall denote this by $\Phi_{\mathbf{k}n}$, so that

$$\boxed{\Phi_{\mathbf{k}n}(\mathbf{r}, \mathbf{x}) = \exp(i\mathbf{k}\cdot\mathbf{r})\, \chi_n(\mathbf{x}).} \tag{4.1.4}$$

We shall assume that the internal states are normalized, so that if χ_m is a second bound state satisfying

$$\mathsf{h}\chi_m = E_m\, \chi_m, \tag{4.1.5}$$

we have

$$\langle \chi_m | \chi_n \rangle = \delta_{mn}, \tag{4.1.6}$$

and since (see, for example, Volume 1, pp. 112–13) $\int \exp[i(\boldsymbol{\kappa}-\mathbf{k})\cdot\mathbf{r}]d\mathbf{r} = (2\pi)^3\, \delta(\mathbf{k}-\boldsymbol{\kappa})$ it follows that

$$\langle \Phi_{\mathbf{k}n} | \Phi_{\boldsymbol{\kappa}m} \rangle = (2\pi)^3\, \delta(\mathbf{k}-\boldsymbol{\kappa})\, \delta_{mn}. \tag{4.1.7}$$

We can therefore define a normalized initial state $\varphi_{\mathbf{k}n}$ by

$$\varphi_{\mathbf{k}n}(\mathbf{r}, \mathbf{x}) = (2\pi)^{-3/2}\, \Phi_{\mathbf{k}n}(\mathbf{r}, \mathbf{x}) = (2\pi)^{-3/2} \exp(i\mathbf{k}\cdot\mathbf{r})\, \chi_n(\mathbf{x}) \tag{4.1.8}$$

so that

$$\langle \varphi_{\mathbf{k}n} | \varphi_{\boldsymbol{\kappa}m} \rangle = \delta(\mathbf{k}-\boldsymbol{\kappa})\, \delta_{mn}. \tag{4.1.9}$$

We shall often denote the quantum numbers $\mathbf{k}n$ by α, so that $\Phi_\alpha = \Phi_{\mathbf{k}n}$, $\Phi_\gamma = \Phi_{\boldsymbol{\kappa}m}$, etc., and (4.1.9) then becomes

$$\langle \varphi_\alpha | \varphi_\gamma \rangle = \delta_{\alpha\gamma}\,. \tag{4.1.10}$$

The Hamiltonian H_0 of the unperturbed system is given by

$$\mathsf{H}_0 = \mathsf{K} + \mathsf{h} \tag{4.1.11}$$

where $\mathsf{K} = -\hbar^2\, \nabla_{\mathbf{r}}^2/2\mu$ is the relative kinetic energy operator. If $\Phi_{\mathbf{k}}(\mathbf{r}) =$

exp $(i\mathbf{k}.\mathbf{r})$ is the plane wave, it follows that

$$\mathsf{K}\Phi_{\mathbf{k}} = E_k\Phi_{\mathbf{k}} \tag{4.1.12}$$

where

$$\boxed{E_k = \hbar^2 k^2/2\mu,} \tag{4.1.13}$$

so that E_k is the relative kinetic energy. Since E_n is the internal energy of the initial system, the quantity

$$\boxed{E_{kn} \equiv E_k+E_n} \tag{4.1.14}$$

is the total energy of the system. Further, $\mathsf{h}\chi_n = E_n\chi_n$, and from (4.1.4)

$$\Phi_{\mathbf{k}n} = \Phi_{\mathbf{k}}\, \chi_n, \tag{4.1.15}$$

hence from (4.1.11) and (4.1.12)

$$\mathsf{H}_0\, \Phi_{\mathbf{k}n} = E_{kn}\, \Phi_{\mathbf{k}n}. \tag{4.1.16}$$

Now $\varphi_\alpha = \varphi_{\mathbf{k}n} = (2\pi)^{-3/2}\, \Phi_{\mathbf{k}n}$, and so we can also denote E_{kn} by E_α, obtaining

$$\mathsf{H}_0\, \varphi_\alpha = E_\alpha\varphi_\alpha. \tag{4.1.17}$$

If the interaction between A and B is V, the total Hamiltonian H is given by

$$\mathsf{H} = \mathsf{H}_0+V, \tag{4.1.18}$$

and the scattering wave function Ψ satisfies the Schrödinger equation

$$\mathsf{H}\Psi = E_{kn}\,\Psi. \tag{4.1.19}$$

In the case of a collision represented by (4.1.1) the final state will have a wave function of the form

$$\Phi_{\mathbf{l}p}(\mathbf{r}, \mathbf{x}) = \exp\,(i\mathbf{l}\cdot\mathbf{r})\, \chi_p(\mathbf{x}), \tag{4.1.20}$$

representing a motion of A and B in which the relative momentum of their centres of mass is $\hbar\mathbf{l}$ and internal state is χ_p. The energy conservation condition demands that

$$E_l+E_p = E_{lp} = E_{kn} = E_k+E_n = E, \tag{4.1.21}$$

where E is the energy of the collision. Hence

$$\hbar^2 l^2/2\mu + E_p = \hbar^2 k^2/2\mu + E_n, \tag{4.1.22}$$

and so

$$l^2 = k^2 + 2\mu(E_n - E_p)/\hbar^2. \tag{4.1.23}$$

In view of these physical considerations, it is reasonable to suppose that the scattering wave function Ψ, in addition to satisfying the wave equation (4.1.19), also satisfies the boundary condition

$$\Psi(\mathbf{r}, \mathbf{x}) \underset{r\to\infty}{\sim} \Phi_{\mathbf{k}n}(\mathbf{r}, \mathbf{x}) + \sum_p{}' f(\hat{\mathbf{r}}|\mathbf{k}n \to p) \frac{\exp(ilr)}{r} \chi_p(\mathbf{x}). \tag{4.1.24}$$

The summation $\sum_p'$ must go over all energetically possible final states; that is to say, all states χ_p for which the right-hand side of (4.1.23) is non-negative, for clearly l must be real. This summation may therefore include states χ_p in the continuum, where one or both of the particles are broken into two or more fragments. Each term of the summation in (4.1.24) represents the scattered wave corresponding to the pth excited state.

If the final relative velocity is v_p, we have $\hbar l = \mu v_p$, and since $\hbar k = \mu v$ where v is the initial relative velocity, we see from (4.1.23) that

$$v_p{}^2 = v^2 + 2(E_n - E_p)/\mu. \tag{4.1.25}$$

Let $d\hat{\mathbf{r}} = d\omega = \sin\theta d\theta d\varphi$ be an element of solid angle in the direction of the radius vector $\mathbf{r}$. The current in the direction of the unit vector $\hat{\mathbf{r}} = \mathbf{r}/r$ is the radial component of the current $\mathbf{J}$ given by (1.2.6), and the reader may easily verify for himself that the current due to the pth scattered wave is given by $v_p|f(\hat{\mathbf{r}}|\mathbf{k}n \to p)|^2 r^{-2}$. It follows that the flux into the solid angle $d\hat{\mathbf{r}}$, which is the number of particles emerging into it per unit time, is $v_p|f(\hat{\mathbf{r}}|\mathbf{k}n \to p)|^2 d\hat{\mathbf{r}}$. The differential cross-section $\sigma(\hat{\mathbf{r}}|\mathbf{k}n \to p)\, d\hat{\mathbf{r}}$ for excitation of the pth excited state is, by definition, this flux divided by the initial flux; this is just v since a plane wave represents a density of one particle per unit volume, and so

$$\boxed{\sigma(\hat{\mathbf{r}}|\mathbf{k}n \to p) = \frac{v_p}{v} |f(\hat{\mathbf{r}}|\mathbf{k}n \to p)|^2.} \tag{4.1.26}$$

Since $\hat{\mathbf{r}}$ is defined by the spherical angles θ, φ relative to the direction Oz of the incident particle A relative to B (B is usually at rest), (4.1.26) can also be written

$$\sigma(\theta, \varphi|\mathbf{k}n \to p) = \frac{v_p}{v}|f(\theta, \varphi|\mathbf{k}n \to p)|^2. \tag{4.1.27}$$

The total cross-section $\sigma(\mathbf{k}n \to p)$ for excitation of the pth excited state is then

$$\sigma(\mathbf{k}n \to p) = \int_0^\pi \sin\theta\, d\theta \int_0^{2\pi} d\varphi\, \sigma(\theta, \varphi|\mathbf{k}n \to p) = \int d\hat{\mathbf{r}}\, \sigma(\hat{\mathbf{r}}|\mathbf{k}n \to p). \tag{4.1.28}$$

4.2. Electron–hydrogen scattering without exchange

The last section dealt with the general case of scattering of two complex particles A and B where change of state of one or both the colliding partners may take place, but where there is no reaction between them. In this section we shall deal with the special case when A is an electron e and B is a hydrogen atom H, and we shall neglect the effect of spin and the Pauli principle for the sake of simplicity of presentation. This approximation is, in fact, perfectly good for high energies.

A collision between e and H where the atom is initially in some state described by the quantum numbers n and finally in a state described by the quantum numbers p is usually denoted by

$$e+H(n) \to e+H(p). \tag{4.2.1}$$

For many purposes it is a good approximation to regard the centre of mass of the hydrogen atom as coinciding with its proton P, so that $\mathbf{r}$ is the position vector of e relative to P. The coordinate $\mathbf{x}$ becomes $\boldsymbol{\rho}$, the position vector of the atomic electron a relative to P, so that the coordinate system is as shown in Fig. 4.1. Since $\gamma = m/M$ (where, as

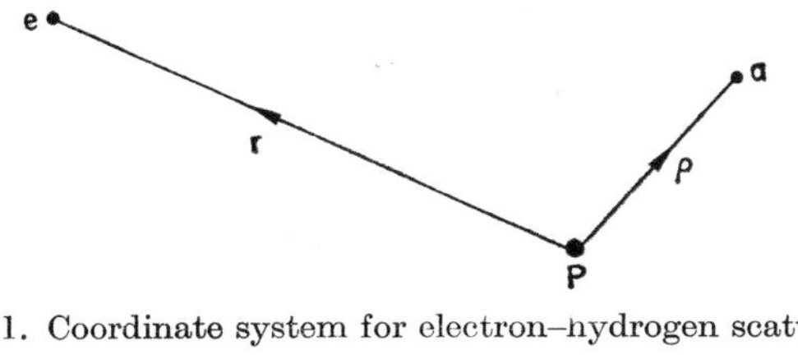

FIG. 4.1. Coordinate system for electron–hydrogen scattering.

usual, m is the mass of an electron and M is the mass of a proton) is very small, the laboratory and centre of mass systems will very nearly coincide in this case, and no distinction need be made between them.

The internal Hamiltonian h is now that of the hydrogen atom, and since $m \ll M$

$$\mathsf{h} = -\frac{\hbar^2}{2m}\nabla_\rho^2 - \frac{e^2}{\kappa_0 \rho}. \tag{4.2.2}$$

The interaction V is given by

$$V = -\frac{e^2}{\kappa_0 r} + \frac{e^2}{\kappa_0 |\mathbf{r}-\boldsymbol{\rho}|} = V(\mathbf{r}, \boldsymbol{\rho}), \tag{4.2.3}$$

and since $\mu = m(m+M)/(2m+M) \simeq m$, the relative kinetic energy operator is

$$\mathsf{K} = -\frac{\hbar^2}{2m}\nabla_\mathbf{r}^2. \tag{4.2.4}$$

The unperturbed Hamiltonian is

$$\mathsf{H}_0 = \mathsf{K} + \mathsf{h}, \tag{4.2.5}$$

and the full Hamiltonian is

$$\mathsf{H} = \mathsf{H}_0 + V. \tag{4.2.6}$$

The internal eigenstates χ_j are the states of the hydrogen atom and satisfy

$$\mathsf{h}\chi_j = E_j \chi_j \tag{4.2.7}$$

where E_j is the energy of the state χ_j. The initial and final states χ_n and χ_p respectively of the atom are particular cases of these, and have energies E_n and E_p; hence

$$\mathsf{h}\chi_n = E_n \chi_n, \quad \mathsf{h}\chi_p = E_p \chi_p. \tag{4.2.8}$$

The scattering wave function $\Psi = \Psi(\mathbf{r}, \boldsymbol{\rho})$ is a solution of the Schrödinger equation

$$\mathsf{H}\Psi = (E_k + E_n)\Psi = E_{kn}\Psi = E\Psi \tag{4.2.9}$$

where E_k is the energy $\hbar^2 k^2/2\mu \simeq \hbar^2 k^2/2m$ of the incident electron, so that $E_{kn} = E_k + E_n$ is the total energy E of the collision which we shall assume is conserved. The initial and final states are given by

$$\Phi_{\mathbf{k}n}(\mathbf{r}, \boldsymbol{\rho}) = \exp(i\mathbf{k}\cdot\mathbf{r})\chi_n(\boldsymbol{\rho}), \quad \Phi_{\mathbf{l}p}(\mathbf{r}, \boldsymbol{\rho}) = \exp(i\mathbf{l}\cdot\boldsymbol{\rho})\chi_p(\boldsymbol{\rho}), \tag{4.2.10}$$

and Ψ obeys the boundary condition (4.1.24) which in this case becomes

$$\Psi(\mathbf{r},\ \boldsymbol{\rho}) \underset{r\to\infty}{\sim} \Phi_{\mathbf{k}n}(\mathbf{r},\ \boldsymbol{\rho})+\sum_{p}{}' f(\hat{\mathbf{r}}|\mathbf{k}n \to p)\,\frac{\exp\,(ilr)}{r}\,\chi_p(\boldsymbol{\rho}). \tag{4.2.11}$$

The conservation of energy implies that l is determined by

$$E_l+E_p = E_k+E_n \tag{4.2.12}$$

where

$$E_k = \frac{\hbar^2 k^2}{2m},\quad E_l = \frac{\hbar^2 l^2}{2m}, \tag{4.2.13}$$

and the summation $\sum_p'$ goes over energetically attainable final states $\Phi_{\mathbf{l}p}$.

In order to solve this scattering problem we first note that the functions χ_j form a complete set. We can therefore expand the wave function $\Psi(\mathbf{r},\ \boldsymbol{\rho})$ in the form

$$\Psi(\mathbf{r},\ \boldsymbol{\rho}) = \sum_j F_j(\mathbf{r})\chi_j(\boldsymbol{\rho}) \tag{4.2.14}$$

where the expansion coefficients F_j must clearly depend upon $\mathbf{r}$. The summation in (4.2.14) must go over *all* j, including states χ_j which are not energetically attainable. If we can calculate the F_j for all j we shall have solved the scattering problem.

Let us substitute for Ψ from (4.2.14) into (4.2.9); if we use (4.2.4)–(4.2.7) we obtain

$$\sum_j \left\{-\frac{\hbar^2}{2m}\nabla_{\mathbf{r}}^2+E_j+V\right\} F_j(\mathbf{r})\ \chi_j(\boldsymbol{\rho}) = \sum_j\ (E_k+E_n)\ F_j(\mathbf{r})\ \chi_j(\boldsymbol{\rho}). \tag{4.2.15}$$

We now multiply (4.2.15) by $\chi_p^*(\boldsymbol{\rho})$ and integrate over the internal atomic coordinate $\boldsymbol{\rho}$; if we use (4.2.12) and (4.2.13) and the fact that the χ_j's form an orthonormal set we find that

$$(\nabla_{\mathbf{r}}^2+l^2)\ F_p(\mathbf{r}) = \sum_j\ U_{pj}(\mathbf{r})\ F_j(\mathbf{r}) \tag{4.2.16}$$

where

$$U_{pj}(\mathbf{r}) = (2m/\hbar^2)\int d\boldsymbol{\rho}\ \chi_p^*(\boldsymbol{\rho})\ V(\mathbf{r},\ \boldsymbol{\rho})\ \chi_j(\boldsymbol{\rho}). \tag{4.2.17}$$

As p varies over all possible values including those for the energetically unattainable states we see that (4.2.16) represents an infinite set of coupled differential equations for the unknown functions $F_p(\mathbf{r})$. When $E_{kn}-E_p > 0$, so that χ_p is energetically attainable, F_p must satisfy the boundary condition

$$F_p(\mathbf{r}) \underset{r\to\infty}{\sim} \delta_{np}\exp\,(i\mathbf{k}\cdot\mathbf{r})+f(\hat{\mathbf{r}}|\mathbf{k}n \to p)\,\frac{\exp\,(ilr)}{r}; \tag{4.2.18}$$

if $E_{kn}-E_p < 0$ we require that F_p decays exponentially as $r \to \infty$. If the F_p satisfy these conditions we see from (4.2.14) that the total wave function Ψ must satisfy the boundary condition (4.2.11) appropriate for the solution of the scattering problem. The problem is now the determination of the F_p's which satisfy (4.2.16) and (4.2.18); this done, (4.2.18) gives us the scattering amplitude $f(\hat{\mathbf{r}}|\mathbf{k}n \to p)$ for the process $\Phi_{\mathbf{k}n} \to \Phi_{\mathbf{l}p}$, and hence the differential cross-section for the collision.

When the amplitudes have been obtained the differential cross-section for excitation of the pth state is given by

$$\sigma(\theta, \varphi|\mathbf{k}n \to p) = (v_p/v)|f(\theta, \varphi|\mathbf{k}n \to p)|^2 \tag{4.2.19}$$

where v and v_p are the initial and final speeds of the electron e. The total cross-section for excitation of the pth state is then

$$\sigma(\mathbf{k}n \to p) = (v_p/v) \int_0^\pi \sin\theta\, d\theta \int_0^{2\pi} d\varphi |f(\theta, \varphi|\mathbf{k}n \to p)|^2. \tag{4.2.20}$$

If there are N atoms $N\sigma(\mathbf{k}n \to p)$ is the number of particles scattered per unit time with excitation of the pth state; in other words, the rate of production of atoms in this state.

The momentum transfer is denoted by $\hbar\mathbf{q}$ as usual, so that

$$\mathbf{q} = \mathbf{k} - \mathbf{l}. \tag{4.2.21}$$

It follows from (4.2.21) that

$$q^2 = k^2 + l^2 - 2kl \cos\theta, \tag{4.2.22}$$

since the scattering angle θ is that between the initial and final momenta $\hbar\mathbf{k}$ and $\hbar\mathbf{l}$. (4.2.22) shows that as θ varies from 0 to π, q varies from $|l-k|$ to $l+k$, and so, since $v_p/v = l/k$, we see from (4.2.22) that (4.2.20) may be written

$$\sigma(\mathbf{k}n \to p) = k^{-2} \int_{|l-k|}^{l+k} q\, dq \int_0^{2\pi} d\varphi |f(q, \varphi|\mathbf{k}n \to p)|^2. \tag{4.2.23}$$

It is not, of course, possible to exactly solve the set of equations (4.2.16), and so in practice we have to make some simplifying approximations, some of which we shall now consider.

Example 4.1. *Scattering of fast electrons by hydrogen atoms.* It is an experimental fact (as well as intuitively reasonable) that scattering diminishes with increasing speed of impact. Hence at high energies we can apply Born's approximation (see Section 2.3); that is to say, we can assume that the effect of the potential is so small that we can approximate the exact scattering state Ψ by the unperturbed initial state $\Phi_{\mathbf{k}n}$. Now if we substitute for U_{pj} from (4.2.17) into (4.2.16) and then use (4.2.14) we see that

$$(\nabla_{\mathbf{r}}^2 + l^2)F_p(\mathbf{r}) = (2m/\hbar^2) \int d\boldsymbol{\rho}\, \chi_p^*(\boldsymbol{\rho})\, V(\mathbf{r}, \boldsymbol{\rho})\, \Psi(\mathbf{r}, \boldsymbol{\rho}). \tag{4.2.24}$$

If on the right-hand side of (4.2.24) we put

$$\Psi(\mathbf{r}, \boldsymbol{\rho}) = \Phi_{\mathbf{k}n}(\mathbf{r}, \boldsymbol{\rho}) = \exp(i\mathbf{kr})\, \chi_n(\boldsymbol{\rho})$$

and then use (4.2.17) with $j = n$ we see that

$$(\nabla_{\mathbf{r}}^2 + l^2) F_p(\mathbf{r}) = U_{pn}(\mathbf{r}) \exp(i\mathbf{k}.\mathbf{r}). \quad (4.2.25)$$

We have to find a solution‡ of (4.2.25) which satisfies the boundary condition (4.2.18).

One such solution is, as we shall show,

$$F_p(\mathbf{r}) = \delta_{np} \exp(i\mathbf{k}\cdot\mathbf{r}) + \int g_l(\mathbf{r}, \mathbf{s})\, U_{pn}(\mathbf{s}) \exp(i\mathbf{k}\cdot\mathbf{s})\, d\mathbf{s} \quad (4.2.26)$$

where $g_l(\mathbf{r}, \mathbf{s})$ is given by (2.1.8) with k replaced by l. Thus, from (4.2.12) and (4.2.13) $l = k$ implies $p = n$ and so $(\nabla_r^2 + l^2)$ operates on the first term of (4.2.26) to give

$$(\nabla_{\mathbf{r}}^2 + l^2)\delta_{np} \exp(i\mathbf{k}\cdot\mathbf{r}) = (\nabla_{\mathbf{r}}^2 + k^2)\delta_{np} \exp(i\mathbf{k}\cdot\mathbf{r}) = 0. \quad (4.2.27)$$

The basic property (2.1.1) of the Green's function also gives (replacing k by l)

$$(\nabla_{\mathbf{r}}^2 + l^2) g_l(\mathbf{r}, \mathbf{s}) = \delta(\mathbf{r} - \mathbf{s}), \quad (4.2.28)$$

and as in Section 2.2 we have, using (4.2.28),

$$\begin{aligned}(\nabla_{\mathbf{r}}^2 + l^2) &\int g_l(\mathbf{r}, \mathbf{s})\, U_{pn}(\mathbf{s}) \exp(i\mathbf{k}\cdot\mathbf{s})\, d\mathbf{s} \\ &= \int (\nabla_{\mathbf{r}}^2 + l^2)\, g_l(\mathbf{r}, \mathbf{s})\, U_{pn}(\mathbf{s}) \exp(i\mathbf{k}\cdot\mathbf{s})\, d\mathbf{s} \\ &= \int \delta(\mathbf{r} - \mathbf{s})\, U_{pn}(\mathbf{s}) \exp(i\mathbf{k}\cdot\mathbf{s})\, d\mathbf{s} \\ &= U_{pn}(\mathbf{r}) \exp(i\mathbf{k}\cdot\mathbf{r}). \end{aligned} \quad (4.2.29)$$

If we operate on (4.2.26) with $(\nabla_{\mathbf{r}}^2 + l^2)$ and use (4.2.27) and (4.2.29) we see that (4.2.25) is satisfied.

We now let $r \to \infty$ in (4.2.26) and apply (2.1.11) with k replaced by l to see that the boundary condition (4.2.18) is also satisfied and that the scattering amplitude is given by

$$f(\hat{\mathbf{r}}|\mathbf{k}n \to p) = -(4\pi)^{-1} \int \exp(-il\hat{\mathbf{r}}\cdot\mathbf{s})\, U_{pn}(\mathbf{s}) \exp(i\mathbf{k}\cdot\mathbf{s})\, d\mathbf{s}. \quad (4.2.30)$$

Now the final wave vector is $\mathbf{l} = l\hat{\mathbf{r}}$, and so (4.2.30) may be rewritten as

$$f(\Phi_{\mathbf{k}n} \to \Phi_{\mathbf{l}p}) = -(4\pi)^{-1} \int \exp(-i\mathbf{l}\cdot\mathbf{r})\, U_{pn}(\mathbf{r}) \exp(i\mathbf{k}\cdot\mathbf{r})\, d\mathbf{r}. \quad (4.2.31)$$

If we substitute for $U_{pn}(\mathbf{r})$ from (4.2.17) into (4.2.31) we obtain

$$\begin{aligned} f(\Phi_{\mathbf{k}n} \to \Phi_{\mathbf{l}p}) = -(m/2\pi\hbar^2) \int d\mathbf{r} \int d\boldsymbol{\rho}\, \exp(-i\mathbf{l}\cdot\mathbf{r})\, \chi_p^*(\boldsymbol{\rho}) \times \\ \times V(\mathbf{r}, \boldsymbol{\rho}) \exp(i\mathbf{k}\cdot\mathbf{r})\, \chi_n(\boldsymbol{\rho}). \end{aligned} \quad (4.2.32)$$

It follows from (4 .2.10) that (4.2.32) may be rewritten in the form

$$f(\Phi_{\mathbf{k}n} \to \Phi_{\mathbf{l}p}) = -(m/2\pi\hbar^2)\, \langle \Phi_{\mathbf{l}p} | \mathsf{V} | \Phi_{\mathbf{k}n} \rangle, \quad (4.2.33)$$

showing how the result (2.3.6) for scattering of a particle by a fixed centre of force generalizes in this case. In the special case when we consider elastic scattering of electrons by hydrogen atoms in their ground state $p = n = 0$. The ground

‡ We assume that the amplitude thus obtained is unique.

state wave function is (Vol. 1, Section 2.3)

$$\chi_0(\boldsymbol{\rho}) = (\pi a_0{}^3)^{-\frac{1}{2}} \exp(-\rho/a_0) \tag{4.2.34}$$

where a_0 is the Bohr radius. Now (4.2.32) may be written as

$$f(\Phi_{\mathbf{k}0} \to \Phi_{\mathbf{l}0}) = -(m/2\pi\hbar^2) \int d\mathbf{r} \exp(i\mathbf{q}\cdot\mathbf{r}) \, V_{00}(\mathbf{r}), \tag{4.2.35}$$

where we have used (4.2.21) and

$$V_{00}(\mathbf{r}) = \int d\boldsymbol{\rho} \, V(\mathbf{r}, \boldsymbol{\rho}) \, |\chi_0(\boldsymbol{\rho})|^2. \tag{4.2.36}$$

We may readily evaluate $V_{00}(\mathbf{r})$ by substituting for χ_0 and V from (4.2.34) and (4.2.3) into (4.2.36). The details are given in Appendix B, and the result is

$$V_{00}(\mathbf{r}) = -\frac{e^2}{\kappa_0}\left(\frac{1}{a_0}+\frac{1}{r}\right) \exp\left(-\frac{2r}{a_0}\right). \tag{4.2.37}$$

The elastic scattering amplitude $f(\Phi_{\mathbf{k}0} \to \Phi_{\mathbf{l}0})$ is obtained by substitution for $V_{00}(\mathbf{r})$ from (4.2.37) into (4.2.35). If we follow the procedure of Section 2.3 [see in particular equations (2.3.5) and (2.3.7)] and then perform an elementary integration we find that

$$f(\Phi_{\mathbf{k}0} \to \Phi_{\mathbf{l}0}) = \frac{2me^2}{\kappa_0 \hbar^2 q} \int_0^\infty \left(\frac{1}{a_0}+\frac{1}{r}\right) \exp\left(-\frac{2r}{a_0}\right) \sin qr \, r \, dr$$

$$= \frac{2me^2(8/a_0{}^2+q^2)}{\kappa_0\hbar^2(4/a_0{}^2+q^2)^2} = f_e(q), \text{ say.} \tag{4.2.38}$$

$f_e(q)$ is the elastic scattering amplitude, and we see that it depends only on the magnitude $\hbar q$ of the momentum transfer. Since the scattering is elastic $k = l$, and so (4.2.22) gives

$$q = 2k \sin \tfrac{1}{2}\theta. \tag{4.2.39}$$

To obtain the total cross-section $\sigma(\mathbf{k}0 \to 0) = \sigma_e$, say, we must apply (4.2.23) when we obtain

$$\sigma_e = 2\pi k^{-2} \int_0^{2k} |f_e(q)|^2 \, q \, dq. \tag{4.2.40}$$

In the high energy limit $k \to \infty$, and we see that σ_e is proportional to k^{-2}, and so is inversely proportional to the energy; the reader may also verify this for himself by substituting for f_e from (4.2.38) into (4.2.40), carrying out the integration and then letting $k \to \infty$.

EXAMPLE 4.2. *Elastic scattering of electrons by hydrogen atoms.* As a first approximation we can neglect excitation of the atom; that is to say, we can replace the exact expression (4.2.14) by the approximate one

$$\Psi(\mathbf{r}, \boldsymbol{\rho}) \simeq F_0(\mathbf{r}) \, \chi_0(\boldsymbol{\rho}) \tag{4.2.41}$$

if we assume that the atom is initially in its ground state χ_0. The elastic-scattering is described by the equation of the set (4.2.24) with $p = 0$; if we substitute (4.2.41) into this we see that

$$(\nabla_{\mathbf{r}}{}^2 + k^2) \, F_0(\mathbf{r}) = U_{00}(\mathbf{r}) \, F_0(\mathbf{r}) \tag{4.2.42}$$

since $k = l$ and U_{00} is given by (4.2.17). The boundary condition on F_0 is given by (4.2.18) with $n = p = 0$ so that

$$F_0(\mathbf{r}) \underset{r\to\infty}{\sim} \exp(i\mathbf{k}\cdot\mathbf{r}) + f_e(\hat{\mathbf{r}}) \frac{\exp(ikr)}{r} \tag{4.2.43}$$

where $f_e(\hat{\mathbf{r}}) = f(\hat{\mathbf{r}}\,|\mathbf{k}0 \to 0)$ is the elastic scattering amplitude. We have thus reduced the problem to that of Chapter 1, viz. scattering of an electron by the central potential given by (4.2.36) or (4.2.37). The effective potential $V_{00}(r)$ is just that for the interaction between the electron and the atom when averaged over all positions of the atomic electron. Physically this means that, even if excitation of the atom is energetically impossible, we have nevertheless neglected the distortion of the charge cloud of the atom by the collision. In other words, we have neglected the *polarization* of the atom by the incoming electron, and for this reason this approximation is usually called the "static approximation". The reader who is interested in the effect of polarization is referred to the book by Drukarev (1965). The problem of determining the elastic scattering from (4.2.42) and (4.2.43) at low energies may be dealt with by the partial wave method discussed in Section 1.4, the potential being given by (4.2.37). If we had taken account of the first excited state in the case when it is energetically attainable, (4.2.16) would reduce to the two coupled differential equations

$$(\nabla_{\mathbf{r}}^2 + k^2)\, F_0(\mathbf{r}) = U_{00}(\mathbf{r})\, F_0(\mathbf{r}) + U_{01}(\mathbf{r})\, F_1(\mathbf{r}), \tag{4.2.44}$$

$$(\nabla_{\mathbf{r}}^2 + l^2)\, F_1(\mathbf{r}) = U_{10}(\mathbf{r})\, F_0(\mathbf{r}) + U_{11}(\mathbf{r})\, F_1(\mathbf{r}). \tag{4.2.45}$$

In practice the neglect of the Pauli principle is a more important source of error, but the solution of (4.2.44) and (4.2.45) (the "two-state" approximation) becomes significant in positron–hydrogen scattering, when the Pauli principle does not apply.

4.3. Rearrangement collisions

We shall now consider collisions of the form

$$A + B \to C + D. \tag{4.3.1}$$

In such a collision an exchange of "elementary particles" takes place, so that the composite particles C and D are different from A and B. The question as to what constitutes an "elementary particle" depends upon the branch of physics with which we are dealing. In atomic and molecular physics and chemistry the elementary particles consist of electrons and atomic nuclei; in nuclear physics the question is more controversial, but until comparatively recently neutrons and protons were regarded as elementary particles, and are still treated as such in problems such as that of the scattering of low-energy neutrons by deuterons.

By far the most important complication arising from a rearrangement collision is the fact that the unperturbed Hamiltonian of the final state is different from that of the initial state. For this reason we introduce the concept of an *arrangement channel*. Initially the system consists of the particles A and B with unperturbed Hamiltonian H_0 and interaction V. This is a particular mode of motion of the elementary particles which make up the system; we shall call this "arrangement channel i", and we shall relabel the unperturbed Hamiltonian H_0 and interaction V as H_i and V_i to emphasize that they describe the initial motion. In the direct collisions described in Section 4.1 the particles may change their states but not their constituent particles, and so remain in the same arrangement channel. If a reaction takes place, however, the product particles C and D are different from A and B, and so the system has gone into a new arrangement channel which we will label f. If we denote the new unperturbed Hamiltonian by H_f and let the interaction between C and D be V_f, we see that the total Hamiltonian H of the system, which consists of the sum of the kinetic energy operators of all the elementary particles and their mutual interactions (after removal of the kinetic energy operator of the centre of mass) must be the same for both arrangement channels, and hence

$$\mathsf{H} = \mathsf{H}_i + V_i = \mathsf{H}_f + V_f. \tag{4.3.2}$$

The initial state must have the form

$$\Phi_{\alpha i}(\mathbf{r}_i, \mathbf{x}_i) = \Phi_{\mathbf{k}ni}(\mathbf{r}_i, \mathbf{x}_i) = \exp(i\mathbf{k}\cdot\mathbf{r}_i)\, \chi_{ni}(\mathbf{x}_i) \tag{4.3.3}$$

where $\mathbf{r}_i$ is the position vector of the centre of mass of A relative to the centre of mass of B. χ_{ni} describes the internal state of A and B, and $\mathbf{x}_i$ stands for the internal coordinates of the colliding systems. The energy of the collision is

$$E = E_{kni} = E_{ki} + E_{ni} \tag{4.3.4}$$

where

$$E_{ki} = \hbar^2 k^2/2\mu_i \tag{4.3.5}$$

is the initial relative kinetic energy, μ_i being the reduced mass of A and B, and E_{ni} is the internal energy of the system in its initial state. It therefore follows that

$$\mathsf{H}_i\, \Phi_{kni} = E_{kni}\, \Phi_{\mathbf{k}ni}. \tag{4.3.6}$$

Similarly the final state must take the form

$$\Phi_{\beta f}(\mathbf{r}_f, \mathbf{x}_f) = \Phi_{\mathbf{l}pf}(\mathbf{r}_f, \mathbf{x}_f) = \exp(i\mathbf{l}\cdot\mathbf{r}_f)\, \chi_{pf}(\mathbf{x}_f) \tag{4.3.7}$$

where $\mathbf{r}_f$ is the position vector of the centre of mass of C relative to the centre of D; its energy is

$$E_{lpf} = E_{lf} + E_{pf} \tag{4.3.8}$$

where

$$E_{lf} = \hbar^2 l^2/2\mu_f, \tag{4.3.9}$$

μ_f is the reduced mass of C and D, and E_{pf} is the internal energy of

C and D. Conservation of energy gives the relation

$$E_{ki} + E_{ni} = E = E_{lf} + E_{pf}, \tag{4.3.10}$$

from which we deduce that

$$l^2 = \mu_f k^2/\mu_i + 2\mu_f(E_{ni} - E_{pf})/\hbar^2. \tag{4.3.11}$$

Only final states are attainable for which the right-hand side is non-negative; there will not be enough energy for other states to be produced.

The scattering state Ψ must satisfy the Schrödinger equation

$$\mathsf{H}\Psi = E\Psi. \tag{4.3.12}$$

Asymptotically as $r_f \to \infty$ it must behave as a sum of scattered waves representing the various possible final states of C and D. There will be no incoming plane wave, since the particles C and D are not present initially, and so

$$\Psi(\mathbf{r}_f, \mathbf{x}_f) \underset{r_f \to \infty}{\sim} \sum_p{}' f(\hat{\mathbf{r}}_f|\mathbf{k}ni \to pf) \frac{\exp(ilr_f)}{r_f} \chi_{pf}(\mathbf{x}_f) \tag{4.3.13}$$

where the sum Σ_p' goes over all energetically attainable states of C and D. $f(\hat{\mathbf{r}}_f|\mathbf{k}ni \to pf)$ is the scattering amplitude corresponding to the production of the state χ_{pf} of C and D from $\Phi_{\mathbf{k}ni}$. The differential cross-section is given by

$$\boxed{\sigma(\hat{\mathbf{r}}_f|\mathbf{k}ni \to pf) = (v_{pf}/v)|f(\hat{\mathbf{r}}_f|\mathbf{k}ni \to pf)|^2} \tag{4.3.14}$$

where v_{pf} is the velocity of C relative to D after the collision; this follows by precisely the same arguments as those of Section 4.1. This can also be written

$$\boxed{\sigma(\Phi_{\mathbf{k}ni} \to \Phi_{\mathbf{l}pf}) = (v_{pf}/v)|f(\Phi_{\mathbf{k}ni} \to \Phi_{\mathbf{l}pf})|^2.} \tag{4.3.15}$$

The relationship between the centre of mass coordinate system and the laboratory system may be obtained by a suitable modification of the arguments and results of Section 3.2. Before the collision the situation is identical to that illustrated in Fig. 3.1, but after the collision the masses m_1 and m_2 must be replaced by the masses m_3 and m_4 of

C and D respectively. Thus v' and v'' of Fig. 3.1(a) refer to the velocities of C and D in the laboratory system, and in Fig. 3.1(b) the velocities v_1 and v_2 after the collision must be replaced by the velocities v_3 and v_4 of C and D in the centre of mass system. Since the velocity which must be imposed on the laboratory system to bring the centre of mass to rest remains v_2, the vector diagram of Fig. 3.2 is altered only by the replacement of v_1 by v_3. As in Section 3.2 we deduce that [cf. (3.2.7)]

$$\tan \theta_l = \frac{\sin \theta}{\cos \theta + \gamma} \tag{4.3.16}$$

where now $\gamma = v_2/v_3$. Also $v_2 = m_1 v/(m_1+m_2)$ and $v_3 = m_4 v_{pf}/(m_3+m_4)$, and (in the case of non-relativistic quantum mechanics) mass is conserved so that $m_1+m_2 = m_3+m_4$; hence we see that

$$\gamma = \frac{v_2}{v_3} = \frac{m_1 \, v}{m_4 \, v_{pf}}. \tag{4.3.17}$$

We can now deduce (3.2.8) and hence the expression (3.2.11) relating the differential cross-sections in the two systems in precisely the same manner as that of Section 3.2.

4.4. Electron–hydrogen scattering with exchange

We have already seen that the problem of direct scattering of electrons by hydrogen atoms is equivalent to the problem of solving (4.2.9) subject to the boundary condition (4.2.11). This naturally suggested using the expansion (4.2.14) for Ψ, and this showed that the direct scattering problem is equivalent to the solution of the set of coupled differential equations (4.2.16) subject to the boundary conditions (4.2.18).

When we consider the possibility of electron exchange we must clearly still solve the full Schrödinger equation

$$\left\{-\frac{\hbar^2}{2m}\nabla_{\mathbf{r}}^2 - \frac{\hbar^2}{2m}\nabla_{\rho}^2 - \frac{e^2}{\kappa_0 \rho} - \frac{e^2}{\kappa_0 r} + \frac{e^2}{\kappa_0 |\mathbf{r}-\boldsymbol{\rho}|}\right\} \Psi(\mathbf{r}, \boldsymbol{\rho}) = E\Psi(\mathbf{r}, \boldsymbol{\rho}) \tag{4.4.1}$$

where $E = E_k + E_n$, but now the asymptotic boundary condition is different. For if we refer to Fig. 4.1 we see that we are concerned with a final state Φ_{lpf} in which electron a is free and travelling to infinity with momentum $\hbar \mathbf{l}$, while electron e is bound to the proton P in the hydrogenic state χ_p. In the initial arrangement channel i, electron e is free and electron a is bound, while in the final arrangement channel f the situation is reversed.‡ For this reason the initial state $\Phi_{\mathbf{k}ni}$ must be given the additional suffix i, to emphasize that it is the electron e which is free with momentum $\hbar\mathbf{k}$.

Due to these considerations we must seek a solution of (4.4.1) subject to the boundary condition

$$\Psi(\mathbf{r}, \boldsymbol{\rho}) \underset{\rho\to\infty}{\sim} \sum_p{}' f(\hat{\boldsymbol{\rho}}|\mathbf{k}ni \to jf) \frac{\exp(il\rho)}{\rho} \chi_p(\mathbf{r}) \tag{4.4.2}$$

where $f(\hat{\boldsymbol{\rho}}|\mathbf{k}ni \to pf)$ is now the scattering amplitude for exchange scattering, l is determined by conservation of energy, and the summation Σ'_p is limited to the energetically attainable states. Instead of expanding Ψ in terms of the basis functions $\chi_j(\boldsymbol{\rho})$ as in (4.2.14) we choose the expansion

$$\Psi(\mathbf{r}, \boldsymbol{\rho}) = \sum_j G_j(\boldsymbol{\rho})\, \chi_j(\mathbf{r}) \tag{4.4.3}$$

so that the boundary condition (4.4.2) is satisfied if

$$G_j(\boldsymbol{\rho}) \underset{\rho\to\infty}{\sim} f(\hat{\boldsymbol{\rho}}|\mathbf{k}ni \to jf) \frac{\exp(il\rho)}{\rho}, \quad (E_{kn} - E_j > 0). \tag{4.4.4}$$

If we substitute for $\Psi(\mathbf{r}, \boldsymbol{\rho})$ from (4.4.3) into (4.4.1) and use the facts that $\chi_j(r)$ satisfies

$$\left(-\frac{\hbar^2}{2m}\nabla_{\mathbf{r}}^2 - \frac{e^2}{\kappa_0 r}\right)\chi_j(\mathbf{r}) = E_j\, \chi_j(\mathbf{r}) \tag{4.4.5}$$

and that l is given by energy conservation we find that

$$\sum_j \left(\nabla_\rho^2 - \frac{2me^2}{\kappa_0\hbar^2|\mathbf{r}-\boldsymbol{\rho}|} + \frac{2me^2}{\kappa_0\hbar^2\rho} + l^2\right) G_j(\boldsymbol{\rho})\, \chi_j(\mathbf{r}) = 0. \tag{4.4.6}$$

On multiplication by $\chi_p^*(\mathbf{r})$ and integration over $\mathbf{r}$ we see that (4.4.6) becomes

$$(\nabla_\rho^2 + l^2)\, G_p(\boldsymbol{\rho}) = \frac{2me^2}{\kappa_0\hbar^2} \sum_j \int \chi_p^*(\mathbf{r}) \left(\frac{1}{|\mathbf{r}-\boldsymbol{\rho}|} - \frac{1}{\rho}\right) \chi_j(\mathbf{r})\, G_j(\boldsymbol{\rho})\, d\mathbf{r}. \tag{4.4.7}$$

Like (4.2.16) the expression (4.4.7) gives rise to an infinite set of simultaneous equations for the unknown functions $G_j(\boldsymbol{\rho})$. In this case, however, not only do the derivatives of the G's appear; the G's also appear inside integrals. We therefore have an infinite set of coupled integro-differential equations.

We shall now use the integral equation approach to obtain an exact expression for the scattering amplitude in terms of the scattering wave

‡ Strictly, the wave function should be antisymmetric in particle coordinates and the two situations cannot be distinguished physically: the effect of antisymmetry is discussed later.

function Ψ. In order to do this we make use of (4.4.3) to rewrite (4.4.7) as

$$(\nabla_\rho^2+l^2)\, G_p(\rho) = \frac{2me^2}{\kappa_0\hbar^2}\int \chi_p^*(\mathbf{r})\left(\frac{1}{|\mathbf{r}-\rho|}-\frac{1}{\rho}\right)\Psi(\mathbf{r},\,\rho)\,d\mathbf{r}. \tag{4.4.8}$$

The final interaction potential V_f is that between the emitted electron a and the atom consisting of P and the captured electron e, and so is given by (see Fig. 4.1)

$$V_f(\mathbf{r},\,\rho) = \frac{e^2}{\kappa_0|\mathbf{r}-\rho|}-\frac{e^2}{\kappa_0\rho}. \tag{4.4.9}$$

We can therefore replace (4.4.8) by the equivalent equation

$$(\nabla_\rho^2+l^2)\, G_p(\rho) = \frac{2m}{\hbar^2}\int \chi_p^*(\mathbf{r})\; V_f(\mathbf{r},\,\rho)\;\Psi(\mathbf{r},\,\rho)\,d\mathbf{r}. \tag{4.4.10}$$

The reader may show in the usual way that if $G_p(\rho)$ satisfies the integral equation

$$G_p(\rho) = \frac{2m}{\hbar^2}\int d\sigma\; g_l(\rho,\,\sigma)\int d\mathbf{r}\;\chi_p^*(\mathbf{r})\;V_f(\mathbf{r},\,\sigma)\;\Psi(\mathbf{r},\,\sigma), \tag{4.4.11}$$

where g_l is the free particle Green's function at energy E_l, then G_p satisfies the integro-differential equation (4.4.10) together with the boundary condition (4.4.4) where the exchange amplitude is given by

$$f(\hat{\rho}|\mathbf{k}ni \to pf) = -\frac{m}{2\pi\hbar^2}\int d\mathbf{r}\int d\sigma\,\exp\,(-il\hat{\rho}\cdot\sigma)\;\chi_p^*(\mathbf{r})\;V_f(\mathbf{r},\,\sigma)\;\Psi(\mathbf{r},\,\sigma). \tag{4.4.12}$$

Since $\mathbf{l} = l\hat{\rho}$ and $\Phi_{\mathbf{l}pf}(\mathbf{r},\,\sigma) = \chi_p(\mathbf{r})\,\exp\,(i\mathbf{l}\cdot\sigma)$ we see that (4.4.12) may be replaced by the equivalent expression

$$\boxed{f(\Phi_{\mathbf{k}ni} \to \Phi_{\mathbf{l}pf}) = -\frac{m}{2\pi\hbar^2}\,\langle\Phi_{\mathbf{l}pf}|V_f|\Psi\rangle.} \tag{4.4.13}$$

EXAMPLE 4.3. *The exchange scattering of fast electrons by hydrogen atoms.* (*The Born–Oppenheimer approximation.*) At high energies there is little scattering, and so we can make the approximation $\Psi \simeq \Phi_{\mathbf{k}ni}$. The expression (4.4.13) now becomes

$$\boxed{f(\Phi_{\mathbf{k}ni} \to \Phi_{\mathbf{l}pf}) = -\frac{m}{2\pi\hbar^2}\,\langle\Phi_{\mathbf{l}pf}|\,V_f|\Phi_{\mathbf{k}ni}\rangle.} \tag{4.4.14}$$

When written out in full using (4.4.9) this is

$$f(\Phi_{\mathbf{k}ni} \to \Phi_{\mathbf{l}pf})$$
$$= -\frac{me^2}{2\pi\hbar^2\kappa_0} \int d\boldsymbol{\rho} \int d\mathbf{r}\, \chi_p^*(\mathbf{r}) \exp(-i\mathbf{l}\cdot\boldsymbol{\rho}) \left(\frac{1}{|\mathbf{r}-\boldsymbol{\rho}|} - \frac{1}{\rho}\right) \chi_n(\boldsymbol{\rho}) \exp(i\mathbf{k}\cdot\mathbf{r}). \tag{4.4.15}$$

Equation (4.4.15) is the Born approximation for the exchange amplitude, and is often referred to as the *Born–Oppenheimer* approximation.

In order to take into account the Pauli principle we must symmetrize (or antisymmetrize) the wave function. If we abbreviate the direct and exchange amplitudes $f(\hat{\mathbf{r}}|\mathbf{k}n \to p)$ and $f(\hat{\boldsymbol{\rho}}|\mathbf{k}ni \to pf)$ by $f(\hat{\mathbf{r}})$ and $g(\hat{\boldsymbol{\rho}})$ respectively, and note the asymptotic forms (4.2.11) and (4.4.2) we see that the symmetrized or antisymmetrized wave function satisfies the boundary conditions

$$\Psi(\mathbf{r}, \boldsymbol{\rho}) \pm \Psi(\boldsymbol{\rho}, \mathbf{r})$$
$$\underset{r\to\infty}{\sim} \exp(i\mathbf{k}\cdot\mathbf{r})\, \chi_n(\boldsymbol{\rho}) + \sum_p{}' \{f(\hat{\mathbf{r}}) \pm g(\hat{\mathbf{r}})\}\, \chi_p(\boldsymbol{\rho}) \frac{\exp(ilr)}{r}, \tag{4.4.16}$$

$$\Psi(\mathbf{r}, \boldsymbol{\rho}) \pm \Psi(\boldsymbol{\rho}, \mathbf{r})$$
$$\underset{r\to\infty}{\sim} \pm\exp(i\mathbf{k}\cdot\boldsymbol{\rho})\, \chi_n(\mathbf{r}) + \sum_p{}' \{\pm f(\hat{\boldsymbol{\rho}}) + g(\hat{\boldsymbol{\rho}}) + g(\hat{\boldsymbol{\rho}})\}\, \chi_p(\mathbf{r}) \frac{\exp(il\rho)}{\rho}. \tag{4.4.17}$$

Equations (4.4.16) and (4.4.17) suggest that the total amplitudes $F_{\pm}(\theta, \varphi)$ are given by

$$F_{\pm}(\theta, \varphi) = f(\theta, \varphi) \pm g(\theta, \varphi), \tag{4.4.18}$$

and in a subsequent volume we shall prove that this is indeed the case. F_+ is the amplitude for excitation of the pth excited state when one electron leaves in the direction specified by the spherical polar angles θ and φ while the other electron remains as the atomic electron. Since the plus sign is associated with the singlet (total spin zero) state and the minus sign is associated with the triplet (total spin one) state of the two electrons, and these occur in the ratio 1:3 (see Section 3.4), we expect the total differential cross-section to be given by

$$\sigma(\theta, \varphi) = (v_p/v)\, \{\tfrac{1}{4}|F_+(\theta, \varphi)|^2 + \tfrac{3}{4}|F_-(\theta, \varphi)|^2\}. \tag{4.4.19}$$

We note from these results that if the exchange cross-section is small, g is small compared with f, and then $F_+ \simeq F_- \simeq f$ which implies that the Pauli principle can be neglected. At high energies the velocity of the incident electron is fast compared with that of the atomic electron and so we should then expect the effect of exchange to be small. This is indeed borne out by a theoretical calculation (Goldberger and Watson, 1964; p. 154).

REFERENCES

DRUKAREV, G. F. (1965) *The Theory of Electron-Atom Collisions,* Academic Press, London.

GOLDBERGER, M. L. and WATSON, K. M. (1964) *Collision Theory,* John Wiley & Sons.

CHAPTER 5

FORMAL SCATTERING THEORY

5.1. Some important integrals

Our object in this chapter will not be to study new physical processes, but rather to develop a new and more formal language for describing the scattering processes we have already met. This formalism was developed from 1949 onwards by Schwinger, Lippmann, Gellman, Goldberger and others. There is a close analogy with the development of vector methods in classical physics. Neither describes new physics or new mathematics, but each provides a powerful and compact notation which, once mastered, enables the physical situations to be described more neatly, more concisely and more clearly. One advantage of such a formalism is that it enables us to express scattering amplitudes as made up of terms each having a physical interpretation (such as multiple scattering, binding effects, etc.) and hence enables us to use our physical intuition to make the approximations necessary to render most actual problems tractable. It also enables us to derive results which would be difficult, if not impossible, to obtain without it.

In order to develop this formalism we must first evaluate certain integrals. Although the evaluation of these integrals is essentially pure mathematics, being an exercise in contour integration, it is not the sort of mathematics which can be relegated to an appendix; for an understanding of the method of calculation is essential to an understanding of the formalism itself. For this reason this section is devoted to the evaluation of integrals of the form

$$I = I(l, \eta, \mathbf{r}, \mathbf{s}) = \int (k^2 \pm l^2 \pm i\eta)^{-1} \exp \{i\mathbf{l}\cdot(\mathbf{r}-\mathbf{s})\}\, d\mathbf{l}. \quad (5.1.1)$$

$\mathbf{r}$ and $\mathbf{s}$ are arbitrary vectors, k is an arbitrary non-negative number, and η is an arbitrary positive number which we can take as small as we like, whilst the integration goes over all values of the vector $\mathbf{l}$.

In order to evaluate I we first put $\mathbf{R} = \mathbf{r}-\mathbf{s}$ so that

$$I = \int (k^2 \pm l^2 \pm i\eta)^{-1} \exp (i\mathbf{l}\cdot\mathbf{R})\, d\mathbf{l}. \quad (5.1.2)$$

We now take spherical polar coordinates with $\mathbf{R}$ as polar axis, so that

l has coordinates (l, θ, φ). Hence (5.1.2) becomes

$$I = \int_0^\infty l^2\, dl \int_0^\pi \sin\theta\, d\theta \int_0^{2\pi} d\varphi (k^2 \pm l^2 \pm i\eta)^{-1} \exp\,(ilR\cos\theta)$$

$$= 2\pi \int_0^\infty l^2\, dl (k^2 \pm l^2 \pm i\eta)^{-1} \int_{-1}^{+1} dt \exp\,(ilRt)$$

where $t = \cos\theta$, and so

$$I = \frac{2\pi}{iR}\{\int_0^\infty (k^2 \pm l^2 \pm i\eta)^{-1} \exp\,(ilR) l\, dl$$
$$- \int_0^\infty (k^2 \pm l^2 \pm i\eta)^{-1} \exp\,(-ilR) l\, dl. \quad (5.1.3)$$

If we replace l by $-l$ in the second integral on the right-hand side of (5.1.3) we find that

$$I = \frac{2\pi}{iR} \int_{-\infty}^{+\infty} (k^2 \pm l^2 \pm i\eta)^{-1} l \exp\,(ilR)\, dl. \quad (5.1.4)$$

In the first place we shall evaluate (5.1.4) when the signs are as follows:

$$I = \frac{2\pi}{iR} \int_{-\infty}^{+\infty} (k^2 - l^2 + i\eta)^{-1} l \exp\,(ilR)\, dl. \quad (5.1.5)$$

The above integral can be evaluated by means of contour integration (see Appendix C). If l is allowed to have complex values, so that $l = l_1 + il_2$, where l_1 and l_2 are real, the integral is taken along the real axis of the complex l-plane, whilst $\exp\,(ilR) = \exp\,(-l_1R)\exp\,(il_2R)$ is exponentially decreasing for $l_2 > 0$; in other words, in the upper half of the complex l-plane. We therefore choose the semicircular contour Γ in the upper half-plane shown in Fig. 5.1, for then the contribution of

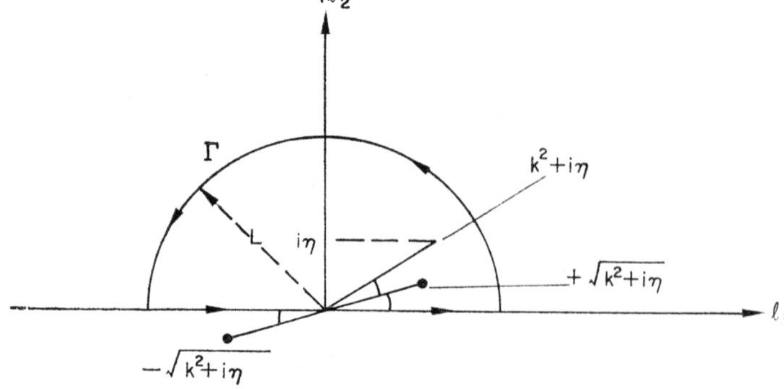

FIG. 5.1. Contour for evaluation of I in (5.1.5).

the circular part of the contour to the integral becomes negligible as the radius L of the contour tends to infinity; and so

$$I = \lim_{L\to\infty} \frac{2\pi}{iR} \int_{\Gamma} (k^2 - l^2 + i\eta)^{-1} l \exp(ilR)\, dl. \tag{5.1.6}$$

The poles of the integrand occur when $l = \pm\sqrt{(k^2+i\eta)}$, where the complex numbers $\pm\sqrt{(k^2+i\eta)}$ are shown in Fig. 5.1. We can see from Fig. 5.1 that the only pole which occurs inside Γ is $l = +\sqrt{(k^2+i\eta)}$, and so if we apply the theorem of residues (see Appendix C) (5.1.6) becomes

$$I = 2\pi i \times \left(\frac{2\pi}{iR}\right) \lim_{l\to\sqrt{(k^2+i\eta)}} [l - \sqrt{(k^2+i\eta)}]\, \frac{l \exp(ilr)}{k^2 - l^2 + i\eta}$$

$$= \left(-\frac{4\pi^2}{R}\right) \lim_{l\to+\sqrt{(k^2+i\eta)}} \frac{l \exp(ilR)}{l + \sqrt{(k^2+i\eta)}},$$

$$\therefore\ I = -\frac{2\pi^2}{R} \exp\{i\sqrt{(k^2+i\eta)}R\}. \tag{5.1.7}$$

If we remember that $\mathbf{R} = \mathbf{r} - \mathbf{s}$, (5.1.1), (5.1.5) and (5.1.7) may be summarized as

$$\int (k^2 - l^2 + i\eta)^{-1} \exp\{i\mathbf{l}\cdot(\mathbf{r}-\mathbf{s})\}\, d\mathbf{l}$$

$$= \frac{2\pi}{i|\mathbf{r}-\mathbf{s}|} \int_{-\infty}^{+\infty} (k^2 - l^2 + i\eta)^{-1} l \exp(il|\mathbf{r}-\mathbf{s}|)\, dl$$

$$= -\frac{2\pi^2}{|\mathbf{r}-\mathbf{s}|} \exp\{i\sqrt{(k^2+i\eta)}|\mathbf{r}-\mathbf{s}|\}. \tag{5.1.8}$$

As a rule we shall be interested in these results in the limit $\eta \to 0+$ (η tends to zero through positive values), and since $+\sqrt{(k^2+i\eta)} \to +k$ as $\eta \to 0+$ we have

$$\lim_{\eta\to 0+} \int (k^2 - l^2 + i\eta)^{-1} \exp\{i\mathbf{l}\cdot(\mathbf{r}-\mathbf{s})\}\, d\mathbf{l}$$

$$= -\frac{2\pi^2}{|\mathbf{r}-\mathbf{s}|} \exp(ik|\mathbf{r}-\mathbf{s}|). \tag{5.1.9}$$

We may note in passing that in the literature integrals of the form

$$\int (k^2 - l^2)^{-1} \exp(i\mathbf{l}\cdot\mathbf{R})\, d\mathbf{l} = \frac{2\pi}{iR} \int_{-\infty}^{+\infty} (k^2 - l^2)^{-1} l \exp(ilR)\, dl \tag{5.1.10}$$

are sometimes met. As it stands, this integral is undefined, since the integrand has a singularity at $l = k$. It may be *defined* as the integral taken along the contour shown in Fig. 5.2 (which is the real axis with two semicircular indentations of radius η at $l = \pm k$) in the limit as

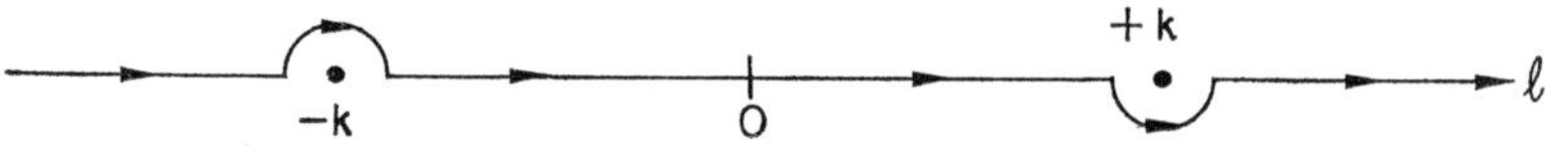

FIG. 5.2. Contour for evaluation of integral in (5.1.10).

$\eta \to 0+$. This integral is evaluated by completing the contour with the same semicircle and the calculation proceeds as above when the result (5.1.9) is again obtained. The two methods of approach are essentially equivalent.

The reader may easily verify for himself that, with $-\eta$ instead of η,

$$\begin{aligned}\lim_{\eta\to+0} \int (k^2-l^2-i\eta)^{-1} \exp\{i\mathbf{l}\cdot(\mathbf{r}-\mathbf{s})\}\, d\mathbf{l} &= \lim_{\eta\to+0} \left\{-\frac{2\pi^2}{|\mathbf{r}-\mathbf{s}|} \exp[-i\sqrt{(k^2-i\eta)}|\mathbf{r}-\mathbf{s}|]\right\} \\ &= -\frac{2\pi^2}{|\mathbf{r}-\mathbf{s}|} \exp(-ik|\mathbf{r}-\mathbf{s}|). \qquad (5.1.11)\end{aligned}$$

It is $-\sqrt{(k^2-i\eta)}$ (which tends to $-k$ as $\eta \to 0+$) which occurs inside the contour this time, and this the only difference compared with the previous calculation.

From (5.1.1) and (5.1.4) we see that, with l^2 instead of $-l^2$,

$$\int (k^2+l^2+i\eta)^{-1} \exp(i\mathbf{l}\cdot\mathbf{R})\, d\mathbf{l} = \frac{2\pi}{iR} \int_{-\infty}^{+\infty} (k^2+l^2+i\eta)^{-1}\, l \exp(ilR)\, dl. \qquad (5.1.12)$$

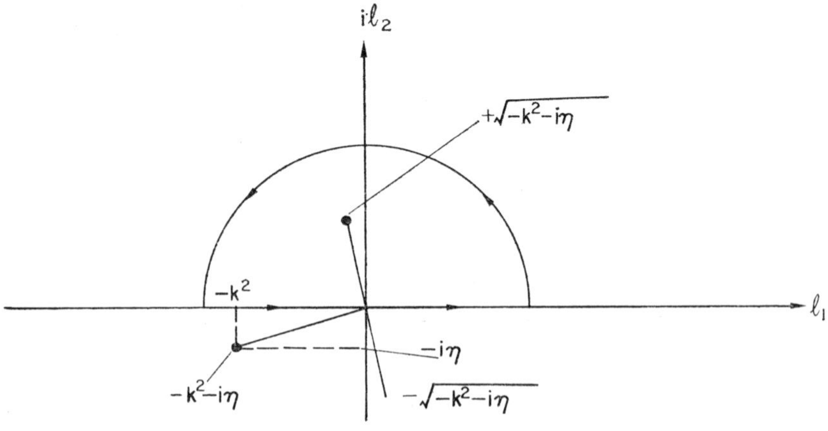

FIG. 5.3. Contour for the evaluation of the integral (5.1.12).

The appropriate contour is shown in Fig. 5.3, and we find that

$$\int (k^2+l^2+i\eta)^{-1} \exp (i\mathbf{l}\cdot\mathbf{R})\, d\mathbf{l}$$

$$= 2\pi i \times \frac{2\pi}{iR} \lim_{l\to\sqrt{(-k^2-i\eta)}} [l-\sqrt{(-k^2-i\eta)}]\, \frac{l \exp (ilR)}{k^2+l^2+i\eta}$$

$$= \frac{2\pi^2}{R} \exp \{i\sqrt{(-k^2-i\eta)}R\}. \qquad (5.1.13)$$

Since $\sqrt{(-k^2-i\eta)} \to ik$ as $\eta \to 0+$ and $\mathbf{R} = \mathbf{r}-\mathbf{s}$, (5.1.13) shows that

$$\lim_{\eta\to 0+} \int (k^2+l^2+i\eta)^{-1} \exp \{i\mathbf{l}\cdot(\mathbf{r}-\mathbf{s})\}\, d\mathbf{l} = \frac{2\pi^2}{|\mathbf{r}-\mathbf{s}|} \exp (-k|\mathbf{r}-\mathbf{s}|). \qquad (5.1.14)$$

The same result is obtained with $i\eta$ replaced by $-i\eta$. The reason for this is that no singularity is introduced into the integrand of (5.1.12) when we put $\eta = 0$; we could, in fact, have put $\eta = 0$ and evaluated the contour integral without going through the limiting process. In the cases when the substitution $\eta = 0$ introduces a singularity, the nature of the limiting process is crucial to the final result.

The above results may be conveniently summarized by

$$\lim_{\eta\to 0+} \int (A-l^2 \pm i\eta)^{-1} \exp \{i\mathbf{l}\cdot(\mathbf{r}-\mathbf{s})\}\, d\mathbf{l}$$

$$= -\frac{2\pi^2}{|\mathbf{r}-\mathbf{s}|} \exp (\pm ik|\mathbf{r}-\mathbf{s}|) \text{ if } A > 0 \qquad (5.1.15)$$

where $k^2 = A$,

and for the alternative choice of sign, corresponding to (5.1.14)

$$\lim_{\eta\to 0+} (A-l^2 \pm i\eta)^{-1} \exp \{i\mathbf{l}\cdot(\mathbf{r}-\mathbf{s})\}\, d\mathbf{l}$$

$$= \int (A-l^2)^{-1} \exp \{i\mathbf{l}\cdot(\mathbf{r}-\mathbf{s})\}\, d\mathbf{l} \qquad (5.1.16)$$

$$= -\frac{2\pi^2}{|\mathbf{r}-\mathbf{s}|} \exp (-k|\mathbf{r}-\mathbf{s}|) \text{ if } A < 0,$$

where $k^2 = -A$.

5.2. Green's operators

Suppose H is a Hermitian operator with orthonormal eigenstates ψ_ν satisfying

$$\mathsf{H}\psi_\nu = E_\nu \psi_\nu, \tag{5.2.1}$$

$$\langle \psi_\mu | \psi_\nu \rangle = \delta_{\mu\nu}. \tag{5.2.2}$$

If they form a complete set, any wave function ψ may be expanded in terms of them according to the formula (e.g. Volume 1, Chapter 3)

$$\psi = \sum_\nu \psi_\nu \langle \psi_\nu | \psi \rangle \tag{5.2.3}$$

where as usual $\langle \psi_\nu | \psi \rangle$ denotes the scalar product of ψ_ν and ψ. If H were the unperturbed Hamiltonian H_0 of Section 4.1, for example, we should have $\psi_\nu = \varphi_\alpha = (2\pi)^{-3/2}\,\Phi_{\mathbf{k}n}$, $E_\nu = E_{\mathbf{k}n}$. In general ν will consist of a set of quantum numbers, and these may take both discrete and continuous values. The summation Σ_ν then implies a sum over discrete values and an integral over continuous values, while $\delta_{\mu\nu}$ becomes a product of Kronecker deltas for the discrete values and Dirac δ-functions for the continuous values.

A function $f(\mathsf{H})$ of H is defined (e.g. Dirac, 1958; or Volume 1, Appendix 4) by the expression

$$f(\mathsf{H})\psi = \sum_\nu f(E_\nu)\, \psi_\nu \langle \psi_\nu | \psi \rangle \tag{5.2.4}$$

for any state ψ. Let us consider the special case when

$$f(\mathsf{H}) = (E - \mathsf{H} + i\varepsilon)^{-1} \tag{5.2.5}$$

where E is real and ε is a positive number which will usually be regarded as small. Then by (5.2.4)

$$(E - \mathsf{H} + i\varepsilon)^{-1}\,\psi = \sum_\nu (E - E_\nu + i\varepsilon)^{-1}\, \psi_\nu \langle \psi_\nu | \psi \rangle. \tag{5.2.6}$$

Since H is Hermitian its eigenvalues E_ν are real and so the expression $E - E_\nu + i\varepsilon$ cannot vanish. This means that the operator $(E - \mathsf{H} + i\varepsilon)^{-1}$ is well defined by (5.2.6). We also note that $|E - E_\nu + i\varepsilon| \geqq \varepsilon$ for all values of ν, and so the series converges if (5.2.4) does. A mathematically rigorous treatment will not be given here but is available elsewhere (Newton, 1966, Section 7.3).

The notation used suggests that $(E - \mathsf{H} + i\varepsilon)^{-1}$ is the inverse of the operator $E - \mathsf{H} + i\varepsilon$. This is indeed the case, for by (5.2.4)

$$
\begin{aligned}
&(E-\mathsf{H}+i\varepsilon)\,(E-\mathsf{H}+i\varepsilon)^{-1}\,\psi \\
&\quad = \sum_{\nu} (E-E_{\nu}+i\varepsilon)\,\psi_{\nu}\,\langle\psi_{\nu}|(E-\mathsf{H}+i\varepsilon)^{-1}|\psi\rangle \\
&\quad = \sum_{\nu} (E-E_{\nu}+i\varepsilon)\,\psi_{\nu}\,\langle\psi_{\nu}|\sum_{\mu} (E-E_{\mu}+i\varepsilon)^{-1}\,\psi_{\mu}\,\langle\psi_{\mu}|\psi\rangle\rangle \\
&\quad = \sum_{\nu}\sum_{\mu} (E-E_{\nu}+i\varepsilon)\,(E-E_{\mu}+i\varepsilon)^{-1}\,\psi_{\nu}\,\langle\psi_{\mu}|\psi\rangle\,\langle\psi_{\nu}|\psi_{\mu}\rangle \\
&\quad = \sum_{\nu}\sum_{\mu} (E-E_{\nu}+i\varepsilon)\,(E-E_{\mu}+i\varepsilon)^{-1}\,\psi_{\nu}\,\langle\psi_{\mu}|\psi\rangle\,\delta_{\mu\nu} \quad \text{[by (2)]} \\
&\quad = \sum_{\nu} \psi_{\nu}\,\langle\psi_{\nu}|\psi\rangle \\
&\quad = \psi,
\end{aligned}
$$

and since this is true for any ψ,

$$(E-\mathsf{H}+i\varepsilon)\,(E-\mathsf{H}+i\varepsilon)^{-1} = \mathsf{I} \tag{5.2.7}$$

where I is the unit operator. The reader may also readily verify that

$$(E-\mathsf{H}+i\varepsilon)^{-1}\,(E-\mathsf{H}+i\varepsilon) = \mathsf{I}, \tag{5.2.8}$$

and so $(E-\mathsf{H}+i\varepsilon)^{-1}$ is the inverse of the operator $E-\mathsf{H}+i\varepsilon$.

The operator $(E-\mathsf{H}-i\varepsilon)^{-1}$ is defined by

$$(E-\mathsf{H}-i\varepsilon)^{-1}\,\psi = \sum_{\nu} (E-E_{\nu}-i\varepsilon)^{-1}\,\psi_{\nu}\,\langle\psi_{\nu}|\psi\rangle, \tag{5.2.9}$$

and similarly may be shown to satisfy

$$(E-\mathsf{H}-i\varepsilon)^{-1}\,(E-\mathsf{H}-i\varepsilon) = (E-\mathsf{H}-i\varepsilon)\,(E-\mathsf{H}-i\varepsilon)^{-1} = \mathsf{I}. \tag{5.2.10}$$

It is important to note that the above definitions depend upon the fact that $\varepsilon \neq 0$. For when $\varepsilon = 0$, the sums or integrals in (5.2.6) or (5.2.9) may not be defined; for example, if E is equal to one of the discrete energy levels of H, one of the terms in the summation over ν will be infinite.

Functions of Hermitian operators have certain properties which we shall note here. If f is a complex-valued function its complex conjugate f^* is defined by

$$f^*(\mathsf{H})\,\psi = \sum_{\nu} f^*(E_{\nu})\,\psi_{\nu}\,\langle\psi_{\nu}|\psi\rangle. \tag{5.2.11}$$

Hence if ψ, ξ are two wave functions,

$$\begin{aligned}\langle\psi|f(\mathsf{H})|\xi\rangle &= \langle\psi|\sum_\nu f(E_\nu)\,\psi_\nu\,\langle\psi_\nu|\xi\rangle\rangle \\ &= \sum_\nu f(E_\nu)\,\langle\psi_\nu|\xi\rangle\,\langle\psi|\psi_\nu\rangle \\ &= \langle\sum_\nu f^*(E_\nu)\,\psi_\nu\,\langle\psi_\nu|\psi\rangle|\xi\rangle \\ &= \langle f^*(\mathsf{H})\,\psi|\xi\rangle,\end{aligned}$$

and so we obtain the important result that

$$[f(\mathsf{H})]^\dagger = f^*(\mathsf{H}) \tag{5.2.12}$$

where † denotes the Hermitian conjugate or adjoint operator (Volume 1, Section 3.7). In particular,

$$[(E-\mathsf{H}\pm i\varepsilon)^{-1}]^\dagger = (E-\mathsf{H}\mp i\varepsilon)^{-1}, \tag{5.2.13}$$

$$[(E-\mathsf{H}_0\pm i\varepsilon)^{-1}]^\dagger = (E-\mathsf{H}_0\mp i\varepsilon)^{-1}. \tag{5.2.14}$$

The results (5.2.7) and (5.2.10) may also be obtained by noting from (5.2.4) that

$$\langle\psi_\nu|f(\mathsf{H})|\psi\rangle = f(E_\nu)\,\langle\psi_\nu|\psi\rangle \tag{5.2.15}$$

so that if $g(\mathsf{H})$ is also a function of H,

$$\begin{aligned}g(\mathsf{H})\,f(\mathsf{H})\,\psi &= g(\mathsf{H})\,\{f(\mathsf{H})\,\psi\} \\ &= \sum_\nu g(E_\nu)\,\psi_\nu\,\langle\psi_\nu|f(\mathsf{H})|\psi\rangle \\ &= \sum_\nu g(E_\nu)\,\psi_\nu\,f(E_\nu)\,\langle\psi_\nu|\psi\rangle;\end{aligned}$$

hence

$$g(\mathsf{H})\,f(\mathsf{H})\,\psi = \sum_\nu g(E_\nu)\,f(E_\nu)\,\psi_\nu\,\langle\psi_\nu|\psi\rangle. \tag{5.2.16}$$

The results

$$(E-\mathsf{H}\pm i\varepsilon)\,(E-\mathsf{H}\pm i\varepsilon)^{-1} = (E-\mathsf{H}\pm i\varepsilon)^{-1}\,(E-\mathsf{H}\pm i\varepsilon) = \mathsf{I} \tag{5.2.17}$$

follow immediately from (5.2.16); we also obtain

$$f(\mathsf{H})\,g(\mathsf{H}) = g(\mathsf{H})\,f(\mathsf{H}), \tag{5.2.18}$$

so that functions of the same Hermitian operator commute.

The operators $(E-H\pm i\varepsilon)^{-1}$ are known as Green's operators for the Hamiltonian H. At the end of any calculation using them we usually allow ε to tend to zero through positive values.

5.3. Green's operator for a free particle

As an example of a Green's operator, let us consider the case when

$$\mathsf{H} = \mathsf{H}_0 = -\frac{\hbar^2}{2\mu}\nabla^2. \tag{5.3.1}$$

As before we use the notation

$$\varphi_{\mathbf{l}}(\mathbf{r}) \equiv (2\pi)^{-3/2} \exp(i\mathbf{l}\cdot\mathbf{r}) \tag{5.3.2}$$

so that the $\varphi_{\mathbf{l}}$ are eigenfunctions of H_0 corresponding to the ψ_ν of Section 5.2, and satisfy

$$\mathsf{H}_0\,\varphi_{\mathbf{l}} = (\hbar^2 l^2/2\mu)\varphi_{\mathbf{l}} = E_l\,\varphi_{\mathbf{l}}. \tag{5.3.3}$$

The orthonormality property (delta function normalization) becomes

$$\langle\varphi_{\mathbf{k}}|\varphi_{\mathbf{l}}\rangle = (2\pi)^{-3}\int \exp[i(\mathbf{l}-\mathbf{k})\cdot\mathbf{r}]\,d\mathbf{r} = \delta(\mathbf{k}-\mathbf{l}) \tag{5.3.4}$$

and clearly (5.3.3) and (5.3.4) correspond to (5.2.1) and (5.2.2) respectively.

If we put

$$E = \hbar^2 k^2/2\mu \tag{5.3.5}$$

and apply (5.2.6) to this case we find that

$$\begin{aligned}(E-\mathsf{H}_0+i\varepsilon)^{-1}\,\psi &= \int d\mathbf{l}\left(E-\frac{\hbar^2 l^2}{2\mu}+i\varepsilon\right)^{-1}\varphi_{\mathbf{l}}\,\langle\varphi_{\mathbf{l}}|\psi\rangle\\ &= \frac{2\mu}{\hbar^2}\int d\mathbf{l}(k^2-l^2+i\eta)^{-1}\,\varphi_{\mathbf{l}}\,\langle\varphi_{\mathbf{l}}|\psi\rangle\end{aligned} \tag{5.3.6}$$

where

$$\eta = 2\mu\varepsilon/\hbar^2 \tag{5.3.7}$$

and the integration goes over all $\mathbf{l}$. Now $(E-\mathsf{H}_0+i\varepsilon)^{-1}\psi$ is a function of the position vector $\mathbf{r}$, and according to (5.3.6) its value $(E-\mathsf{H}_0+i\varepsilon)^{-1}$ $\psi(\mathbf{r})$ at $\mathbf{r}$ is given by

$$(E-\mathsf{H}_0+i\varepsilon)^{-1}\,\psi(\mathbf{r}) = \frac{2\mu}{\hbar^2}\int d\mathbf{l}(k^2-l^2+i\eta)^{-1}\,\varphi_{\mathbf{l}}(\mathbf{r})\,\langle\varphi_{\mathbf{l}}|\psi\rangle; \tag{5.3.8}$$

but

$$\begin{aligned}&\int d\mathbf{l}(k^2-l^2+i\eta)^{-1}\,\varphi_{\mathbf{l}}(\mathbf{r})\,\langle\varphi_{\mathbf{l}}|\psi\rangle\\ &\quad= \int d\mathbf{l}(k^2-l^2+i\eta)^{-1}\,\varphi_{\mathbf{l}}(\mathbf{r})\int d\mathbf{s}\,\varphi_{\mathbf{l}}^*(\mathbf{s})\,\psi(\mathbf{s})\\ &\quad= \int\left\{\int(k^2-l^2+i\eta)^{-1}\,\varphi_{\mathbf{l}}(\mathbf{r})\,\varphi_{\mathbf{l}}^*(\mathbf{s})\,d\mathbf{l}\right\}\psi(\mathbf{s})\,d\mathbf{s}\end{aligned} \tag{5.3.9}$$

if we assume that the order of integration can be interchanged. Equations (5.3.8) and (5.3.9) show that

$$(E - \mathsf{H}_0 + i\varepsilon)^{-1}\, \psi(\mathbf{r}) = \int G_\eta^+(\mathbf{r}, \mathbf{s})\, \psi(\mathbf{s})\, d\mathbf{s} \tag{5.3.10}$$

where

$$G_\eta^+(\mathbf{r}, \mathbf{s}) = \frac{2\mu}{\hbar^2} \int (k^2 - l^2 + i\eta)^{-1}\, \varphi_\mathbf{l}(\mathbf{r})\, \varphi_\mathbf{l}^*(\mathbf{s})\, d\mathbf{l}. \tag{5.3.11}$$

It is easy to evaluate G_η^+, for if we substitute for $\varphi_\mathbf{l}$ from (5.3.2) into (5.3.11) we have

$$G_\eta^+(\mathbf{r}, \mathbf{s}) = \frac{2\mu}{\hbar^2} (2\pi)^{-3} \int (k^2 - l^2 + i\eta)^{-1} \exp\{i\mathbf{l}\cdot(\mathbf{r} - \mathbf{s})\}\, d\mathbf{l}. \tag{5.3.12}$$

The integral appearing in (5.3.12) is one of those evaluated in Section 5.1, and its value is given by (5.1.8); substitution of this value into (5.3.12) gives

$$G_\eta^+(\mathbf{r}, \mathbf{s}) = -\frac{\mu}{2\pi\hbar^2|\mathbf{r} - \mathbf{s}|} \exp\{i\sqrt{(k^2 + i\eta)}|\mathbf{r} - \mathbf{s}|\}. \tag{5.3.13}$$

For reasons that should become clear later on, we usually require the value of G_η^+ in the limit $\eta \to 0+$; (5.3.13) then gives us

$$\lim_{\eta \to 0+} G_\eta^+(\mathbf{r}, \mathbf{s}) = -\frac{\mu}{2\pi\hbar^2|\mathbf{r} - \mathbf{s}|} \exp(ik|\mathbf{r} - \mathbf{s}|) = G_0^+(\mathbf{r}, \mathbf{s}) \text{ say,} \tag{5.3.14}$$

since $\sqrt{(k^2 + i\eta)} \to k$ as $\eta \to 0+$. If the limit as $\varepsilon \to 0+$ (that is, $\eta \to 0+$) can be taken under the integral sign in (5.3.10) we get

$$\lim_{\varepsilon \to 0+} (E - \mathsf{H}_0 + i\varepsilon)^{-1}\, \psi = \mathsf{G}_0^+\, \psi \tag{5.3.15}$$

where G_0^+ is the integral operator with kernel given by (5.3.14).

When $i\varepsilon$ is replaced by $-i\varepsilon$, $i\eta$ is replaced by $-i\eta$ throughout, and in particular in (5.3.12). Hence we must use (5.1.11) and obtain the result

$$(E - \mathsf{H}_0 - i\varepsilon)^{-1}\, \psi(\mathbf{r}) = \int G_\eta^-(\mathbf{r}, \mathbf{s})\, \psi(\mathbf{s})\, d\mathbf{s} \tag{5.3.16}$$

where

$$G_\eta^-(\mathbf{r}, \mathbf{s}) = -\frac{\mu}{2\pi\hbar^2|\mathbf{r} - \mathbf{s}|} \exp\{-i\sqrt{(k^2 - i\eta)}|\mathbf{r} - \mathbf{s}|\}. \tag{5.3.17}$$

Thus (5.3.15) is replaced by

$$\lim_{\varepsilon \to 0+} (E - \mathsf{H}_0 - i\varepsilon)^{-1}\, \psi = \mathsf{G}_0^-\, \psi \tag{5.3.18}$$

where

$$G_0^-(\mathbf{r},\ \mathbf{s}) = -\frac{\mu}{2\pi\hbar^2|\mathbf{r}-\mathbf{s}|} \exp\,(-ik|\mathbf{r}-\mathbf{s}|). \qquad (5.3.19)$$

We might be tempted to conclude from (5.3.15) that

$$(E-\mathsf{H}_0)^{-1}\,\psi = \mathsf{G}_0^+\,\psi.$$

The same argument applied to (5.3.18) would lead to

$$(E-\mathsf{H}_0)^{-1}\,\psi = \mathsf{G}_0^-\,\psi,$$

but it is clear from (5.3.14) and (5.3.19) that G_0^+ and G_0^- are quite distinct integral operators. The reason for this discrepancy is that the inverse $(E-\mathsf{H}_0)^{-1}$ is not, in fact, always defined. For suppose it were; we already know that

$$(E-\mathsf{H}_0)\varphi_\mathbf{k} = 0. \qquad (5.3.20)$$

Operating on the left with $(E-\mathsf{H}_0)^{-1}$ gives

$$(E-\mathsf{H}_0)^{-1}\,(E-\mathsf{H}_0)\varphi_\mathbf{k} = 0;$$

but $(E-\mathsf{H}_0)^{-1}\,(E-\mathsf{H}_0) = \mathsf{I}$, and since $\varphi_\mathbf{k} \neq 0$ there is a contradiction. One avoids this mistake by recalling that

$$(E-\mathsf{H}_0)^{-1}\,\psi = \int (E-E_k)^{-1}\,\varphi_\mathbf{k}\,\langle\varphi_\mathbf{k}|\psi\rangle\,d\mathbf{k}, \qquad (5.3.21)$$

and to avoid the singularity in the integrand we must give E a *complex value*; that is, replace E by $E \pm i\varepsilon$, as shown above. The limit $\varepsilon \to 0+$ must be taken with care. The problem does not arise when E is *negative*, for then there is no singularity in (5.3.21); we must use (5.1.14), and as pointed out at the end of Section 5.1, the same result is obtained whether we use $i\varepsilon$, $-i\varepsilon$, and let $\varepsilon \to 0+$, or put $\varepsilon = 0$ right from the beginning.

By using (2.1.1) and (2.1.8) the reader may show that

$$\left(\frac{\hbar^2}{2\mu}\,\nabla_\mathbf{r}^2 + E\right)\int G_0^+(\mathbf{r},\ \mathbf{s})\,\psi(\mathbf{s})\,d\mathbf{s} = \psi(\mathbf{r}). \qquad (5.3.22)$$

Now ψ is an arbitrary function, and so (5.3.1) gives

$$(E-\mathsf{H}_0)\mathsf{G}_0^+ = \mathsf{I}. \qquad (5.3.23)$$

Thus G_0^+ is a right inverse of $E-\mathsf{H}_0$. (The reasoning which led to the contradiction above assumed that $E-\mathsf{H}_0$ had a left inverse.) Similarly one may prove that

$$(E-\mathsf{H}_0)\mathsf{G}_0^- = \mathsf{I}. \qquad (5.3.24)$$

Finally, it should be noted that G_0^+ is precisely the same as the integral operator G_0 discussed in Section 2.4. In that section we were not concerned with the operator G_0^-, and for that reason did not introduce the plus sign there. Where no ambiguity is likely to arise, we shall continue to drop the sign.

5.4. The Schwinger–Lippmann equation

A collision process, whether it is the scattering of a particle by a centre of force as described in Chapter 1 or the collision of two complex particles described in Chapter 4, is always described by a Hamiltonian H consisting of two parts, namely H_0, the Hamiltonian of the unperturbed system, and V, the interaction, so that

$$\mathsf{H} = \mathsf{H}_0 + V. \tag{5.4.1}$$

Let Φ_α be the initial unperturbed state so that, for example, for potential scattering (that is, scattering of a particle by a centre of force) $\Phi_\alpha(\mathbf{r})$ is given by

$$\Phi_\alpha(\mathbf{r}) = \exp(i\mathbf{k}\cdot\mathbf{r}), \tag{5.4.2}$$

and for collisions of complex particles

$$\Phi_\alpha(\mathbf{r}, \mathbf{x}) = \Phi_{\mathbf{k}n}(\mathbf{r}, \mathbf{x}) = \exp(i\mathbf{k}\cdot\mathbf{r})\,\chi_n(\mathbf{x}). \tag{5.4.3}$$

Φ_α, which is most conveniently left unnormalized, then satisfies

$$\mathsf{H}_0\,\Phi_\alpha = E_\alpha\,\Phi_\alpha = E\,\Phi_\alpha \tag{5.4.4}$$

and differs from the eigenstates in Section 5.2 only by a factor $(2\pi)^{3/2}$.

We now define a state Ψ^+ by the equation

$$\boxed{\Psi^+ \equiv \Phi_\alpha + (E-\mathsf{H}+i\varepsilon)^{-1}\,V\,\Phi_\alpha.} \tag{5.4.5}$$

Ψ^+ is thus fixed once the unperturbed state Φ_α is fixed, for this then determines E. The operator $(E-\mathsf{H}+i\varepsilon)^{-1}$ was defined in Section 5.2, and is the inverse of $E-\mathsf{H}+i\varepsilon$. Hence if we operate on (5.4.5) to the left with $E-\mathsf{H}+i\varepsilon$ and use (5.4.1) we have

$$(E-\mathsf{H}+i\varepsilon)\,\Psi^+ = (E-\mathsf{H}_0-V+i\varepsilon)\,\Phi_\alpha + V\,\Phi_\alpha.$$

But Φ_α is the unperturbed state, so that $\mathsf{H}_0\,\Phi_\alpha = E\,\Phi_\alpha$. Hence

$$(E-\mathsf{H})\,\Psi^+ = i\varepsilon(\Phi_\alpha - \Psi^+). \tag{5.4.6}$$

It follows from (5.4.6) that $(E-\mathsf{H})\,\Psi^+ \to 0$ as $\varepsilon \to 0+$, and so Ψ^+ can be made to satisfy the Schrödinger equation

$$\mathsf{H}\,\Psi^+ = E\,\Psi^+ \tag{5.4.7}$$

as nearly as we like. We shall assume that, as $\varepsilon \to 0+$, Ψ^+ tends to a solution of (5.4.7); this is usually true for potentials which are sufficiently well behaved and tend to zero sufficiently fast as $r \to \infty$. (It is not true, however, for the Coulomb potential, see Section 2.5.)

Let us now rearrange (5.4.6) in the form

$$(E - \mathsf{H}_0 + i\varepsilon)\,(\Psi^+ - \Phi_\alpha) = V\,\Psi^+$$

and operate to the left with $(E - \mathsf{H}_0 + i\varepsilon)^{-1}$. We obtain

$$\boxed{\Psi^+ = \Phi_\alpha + (E - \mathsf{H}_0 + i\varepsilon)^{-1}\,V\,\Psi^+.} \tag{5.4.8}$$

This is known as the Schwinger–Lippmann equation, having first been introduced by Lippmann and Schwinger (1950). Its importance at this stage is that it enables us to show that the state Ψ^+ defined by (5.4.5) is a solution of (5.4.7) which behaves asymptotically like an outgoing wave, so that Ψ^+ is a solution which describes the physical situation. In this section we shall demonstrate this for the special case of potential scattering, and then in the next section we shall discuss the meaning of the Schwinger–Lippmann equation for the general processes described in Section 4.1.

EXAMPLE 5.1. *Schwinger–Lippmann equation for potential scattering.* In functional notation (8) can be written

$$\Psi^+(\mathbf{r}) = \Phi_\alpha(\mathbf{r}) + (E - \mathsf{H}_0 + i\varepsilon)^{-1} V \Psi^+(\mathbf{r}). \tag{5.4.9}$$

If we now use (5.4.2) and (5.3.10) (with ψ replaced by $V\Psi^+$) we see that

$$\Psi^+(\mathbf{r}) = \exp(i\mathbf{k}\cdot\mathbf{r}) + \int G_\eta^+(\mathbf{r}, \mathbf{s})\, V\Psi^+(\mathbf{s})\, d\mathbf{s}. \tag{5.4.10}$$

Now V is a multiplicative operator, so that $V\Psi^+(\mathbf{s}) = V(\mathbf{s})\Psi^+(\mathbf{s})$, and $G_\eta^+(\mathbf{r}, \mathbf{s})$ is given by (5.3.13). If we let $\eta \to 0+$ and assume that the limit can be taken under the integral sign so that (5.3.14) can be applied, we find that (5.4.10) becomes

$$\Psi^+(\mathbf{r}) = \exp(i\mathbf{k}\cdot\mathbf{r}) - \frac{\mu}{2\pi\hbar^2} \int \frac{\exp(ik|\mathbf{r}-\mathbf{s}|)}{|\mathbf{r}-\mathbf{s}|}\, V(\mathbf{s})\, \Psi^+(\mathbf{s})\, d\mathbf{s} \tag{5.4.11}$$

which is identical with the integral equation (2.3.1) when Ψ is replaced by Ψ^+. Thus as we saw in Section 2.2, Ψ^+ (or Ψ) is just the solution of the Schrödinger equation which behaves asymptotically as an outgoing wave, confirming the remarks following (5.4.8).

A second state Ψ^- may be defined by the equation

$$\boxed{\Psi^- = \Phi_\alpha + (E - \mathsf{H} - i\varepsilon)^{-1} V\Phi_\alpha.} \tag{5.4.12}$$

The reader may verify that Ψ^- satisfies the equation

$$\Psi^- = \Phi_\alpha + (E - \mathsf{H}_0 - i\varepsilon)^{-1} V \Psi^- \tag{5.4.13}$$

and for potential scattering he may use equations (5.3.16) to (5.3.19) to show that (5.4.13) is equivalent to an integral equation whose solution behaves asymptotically as a plane wave plus an incoming wave; for the asymptotic form of $G_0^-(\mathbf{r}, \mathbf{s})$ is obtained from (2.1.11) by replacement of k by $-k$, giving the desired incoming wave.

The "Möller operators" (or "Wave operators") Ω^+ and Ω^- are defined by

$$\Omega^\pm = 1 + (E - \mathsf{H}_0 + i\varepsilon)^{-1} V. \tag{5.4.14}$$

Thus (5.4.5) and (5.4.12) may be rewritten

$$\Psi^+ = \Omega^+ \Phi_\alpha, \tag{5.4.15}$$

$$\Psi^- = \Omega^- \Phi_\alpha. \tag{5.4.16}$$

In other words, Ω^+ is just the operator which transforms the unperturbed initial state Φ_α into the scattering state Ψ^+. Ω^- transforms Φ_α into the unphysical state Ψ^-, whose significance lies in the number of important results in which it occurs.

5.5. Formal theory of direct collisions

We shall now deal with the formal theory of direct collisions. By a direct collision we shall mean one of the form $A + B \rightarrow A + B$, although the state of one or both particles may have changed.

Our first task is to examine the nature of the Green's operator $(E - \mathsf{H}_0 + i\varepsilon)^{-1}$ for a system of the form $A + B$ as described in Section 4.1. The normalized eigenstates φ_β of the unperturbed Hamiltonian H_0 have the form

$$\varphi_\beta(\mathbf{r}, \mathbf{x}) = \varphi_{\mathbf{l}p}(\mathbf{r}, \mathbf{x}) = (2\pi)^{-3/2} \Phi_{\mathbf{l}p}(\mathbf{r}, \mathbf{x}) = (2\pi)^{-3/2} \exp(i\mathbf{l}\cdot\mathbf{r})\, \chi_p(\mathbf{x}). \tag{5.5.1}$$

As we saw in Section 4.1 the energy E_β of the state φ_β is $E_{lp} = \hbar^2 l^2/2\mu + E_p$, and the energy E of the collision is the energy $E_{kn} = \hbar^2 k^2/2\mu + E_n$ of the initial state $\Phi_{\mathbf{k}n}$. Hence according to the definition (5.2.6) with H replaced by H_0 the value of $(E - \mathsf{H}_0 + i\varepsilon)^{-1}\psi$ at $\mathbf{r}, \mathbf{x}$ is

$$(E-\mathsf{H}_0+i\varepsilon)^{-1}\,\psi(\mathbf{r},\mathbf{x})$$

$$= \sum_{\beta} (E-E_\beta+i\varepsilon)^{-1}\,\varphi_\beta(\mathbf{r},\mathbf{x})\,\langle\varphi_\beta|\psi\rangle$$

$$= \sum_{p} \int d\mathbf{l}\left(\frac{\hbar^2 k^2}{2\mu}+E_n-\frac{\hbar^2 l^2}{2\mu}-E_p+i\varepsilon\right)^{-1}(2\pi)^{-3/2}\exp(i\mathbf{l}\cdot\mathbf{r})\,\chi_p(\mathbf{x})\times$$

$$\times\int d\mathbf{s}\int d\mathbf{y}\,(2\pi)^{-3/2}\exp(-i\mathbf{l}\cdot\mathbf{s})\,\chi_p^*(\mathbf{y})\,\psi(\mathbf{s},\mathbf{y})$$

where the integration $\int d\mathbf{y}$ goes over the internal coordinates $\mathbf{y}$ of A and B and ψ is some wave function; this may be rearranged to give

$$(E-\mathsf{H}_0+i\varepsilon)^{-1}\,\psi(\mathbf{r},\mathbf{x})$$

$$= \int d\mathbf{s}\int d\mathbf{y}\left\{\sum_{p}\frac{\mu}{4\pi^3\hbar^2}\,\chi_p(\mathbf{x})\,\chi_p^*(\mathbf{y})\times\right.$$

$$\left.\times\int d\mathbf{l}\left[k^2+\frac{2\mu}{\hbar^2}(E_n-E_p)-l^2+i\eta\right]^{-1}\exp[i\mathbf{l}\cdot(\mathbf{r}-\mathbf{s})]\right\}\psi(\mathbf{s},\mathbf{y}) \qquad (5.5.2)$$

where $\eta = 2\mu\varepsilon/\hbar^2$.

We can perform the integration over $\mathbf{l}$ by making use of the results (5.1.15) and (5.1.16) with

$$A = k^2+\frac{2\mu}{\hbar^2}(E_n-E_p).$$

The summation $\sum_p$ must be divided into two parts; a summation $\sum_p'$ over those energetically possible values of p for which $A > 0$, and a summation $\sum_p''$ over the remaining values of p for which $A < 0$. The expression (5.1.15) applies to terms in the first sum, while (5.1.16) applies to terms of the second sum. We therefore find that

$$(E-\mathsf{H}_0+i\varepsilon)^{-1}\,\psi(\mathbf{r},\mathbf{x})$$

$$= \int d\mathbf{s}\int d\mathbf{y}\left(-\frac{\mu}{2\pi\hbar^2}\right)\left\{\sum_{p}{}'\,\chi_p(\mathbf{x})\,\chi_p^*(\mathbf{y})\,\frac{\exp(ik_p|\mathbf{r}-\mathbf{s}|)}{|\mathbf{r}-\mathbf{s}|}\right.$$

$$\left.+\sum_{p}{}''\,\chi_p(\mathbf{x})\,\chi_p^*(\mathbf{y})\,\frac{\exp(-k_p|\mathbf{r}-\mathbf{s}|)}{|\mathbf{r}-\mathbf{s}|}\right\}\psi(\mathbf{s},\mathbf{y}) \qquad (5.5.3)$$

where k_p is determined from the conditions

$$k_p{}^2 = \pm[k^2+2\mu(E_n-E_p)/\hbar^2], \qquad (5.5.4)$$

the plus-and-minus signs being determined according to whether $A > 0$ or $A < 0$.

Now we can rewrite (5.5.3) in the form

$$(E-\mathsf{H}_0+i\varepsilon)^{-1}\,\psi(\mathbf{r},\,\mathbf{x}) = \int d\mathbf{s}\int d\mathbf{y}\; G_0^+(\mathbf{r},\,\mathbf{x};\,\mathbf{s},\,\mathbf{y})\,\psi(\mathbf{s},\,\mathbf{y}) \tag{5.5.5}$$

where the function G_0^+ is given by

$$G_0^+(\mathbf{r},\,\mathbf{x};\,\mathbf{s},\,\mathbf{y}) \equiv \left(-\frac{\mu}{2\pi\hbar^2}\right)\left\{\sum_p{}'\,\chi_p(\mathbf{x})\,\chi_p{}^*(\mathbf{y})\,\frac{\exp\,(ik_p|\mathbf{r}-\mathbf{s}|)}{|\mathbf{r}-\mathbf{s}|} + \sum_p{}''\,\chi_p(\mathbf{x})\,\chi_p{}^*(\mathbf{y})\,\frac{\exp\,(-k_p|\mathbf{r}-\mathbf{s}|)}{|\mathbf{r}-\mathbf{s}|}\right\}. \tag{5.5.6}$$

This shows that $(E-\mathsf{H}_0+i\varepsilon)^{-1}$ is an integral operator with kernel $G_0^+(\mathbf{r},\,\mathbf{x};\,\mathbf{s},\,\mathbf{y})$.

If we use (2.1.8) and (2.1.11) with k replaced by k_p we see that

$$\frac{\exp\,(ik_p|\mathbf{r}-\mathbf{s}|)}{|\mathbf{r}-\mathbf{s}|}\;\underset{r\to\infty}{\sim}\;\frac{\exp\,(ik_pr-ik_p\hat{\mathbf{r}}\cdot\mathbf{s})}{r} \tag{5.5.7}$$

and so (5.5.3) gives the asymptotic form

$$(E-\mathsf{H}_0+i\varepsilon)^{-1}\,\psi(\mathbf{r},\,\mathbf{x}) \underset{r\to\infty}{\sim} \sum_p{}'\left(-\frac{\mu}{2\pi\hbar^2}\right)\frac{\exp(ik_pr)}{r}\,\chi_p(\mathbf{x}) \times\int d\mathbf{s}\int d\mathbf{y}\,\exp\,(-ik_p\hat{\mathbf{r}}\cdot\mathbf{s})\,\chi_p{}^*(\mathbf{y})\,\psi(\mathbf{s},\,\mathbf{y}) \tag{5.5.8}$$

which shows that $(E-\mathsf{H}_0+i\varepsilon)^{-1}\,\psi$ gives rise to outgoing waves. The sum $\sum_p''$ does not appear, since these terms decay exponentially as $r\to\infty$.

We now return to the Schwinger–Lippmann equation (5.4.8). As a functional relationship this can be written

$$\Psi^+(\mathbf{r},\,\mathbf{x}) = \Phi_\alpha(\mathbf{r},\,\mathbf{x})+(E-\mathsf{H}_0+i\varepsilon)^{-1}\,V\,\Psi^+(\mathbf{r},\,\mathbf{x}). \tag{5.5.9}$$

The second term on the right-hand side of (5.5.9) is given by (5.5.5) with ψ replaced by $V\Psi^+$. Since $V\Psi^+(\mathbf{s},\,\mathbf{y}) = V(\mathbf{s},\,\mathbf{y})\,\Psi^+(\mathbf{s},\,\mathbf{y})$, and $\Phi_\alpha(\mathbf{r},\,\mathbf{x}) = \exp\,(i\mathbf{k}\cdot\mathbf{r})\,\chi_n(\mathbf{x})$, (5.5.9) can be written

$$\Psi^+(\mathbf{r},\,\mathbf{x}) = \exp\,(i\mathbf{k}\cdot\mathbf{r})\,\chi_n(\mathbf{x}) +\int d\mathbf{s}\int d\mathbf{y}\;G_0^+(\mathbf{r},\,\mathbf{x};\,\mathbf{s},\,\mathbf{y})\,V(\mathbf{s},\,\mathbf{y})\,\Psi^+(\mathbf{s},\,\mathbf{y}). \tag{5.5.10}$$

Thus, for particles with structure the Schwinger–Lippmann equation is again an integral equation.

We now let $r \to \infty$ in (5.5.9) and use (5.5.8) with $\psi = V\Psi^+$ to obtain the asymptotic form

$$\Psi^+(\mathbf{r}, \mathbf{x}) \underset{r\to\infty}{\sim} \Phi_{\mathbf{k}n}(\mathbf{r}, \mathbf{x}) + \sum_p{}' \left(-\frac{\mu}{2\pi\hbar^2}\right) \frac{\exp(ik_p r)}{r} \chi_p(\mathbf{x}) \times$$
$$\times \int d\mathbf{s} \int d\mathbf{y} \exp(-ik_p \hat{\mathbf{r}}\cdot\mathbf{s})\, \chi_p^*(\mathbf{y})\, V(\mathbf{s}, \mathbf{y})\, \Psi^+(\mathbf{s}, \mathbf{y}). \tag{5.5.11}$$

The magnitude l of the final wave number $\mathbf{l}$ is given by (4.1.23), and comparison of this with (5.5.4), in which we must take the positive sign, shows us that $l = k_p$. (5.5.11) can therefore be rewritten

$$\Psi^+(\mathbf{r}, \mathbf{x}) \underset{r\to\infty}{\sim} \Phi_{\mathbf{k}n}(\mathbf{r}, \mathbf{x}) + \sum_p{}' f(\hat{\mathbf{r}}|\mathbf{k}n \to p) \frac{\exp(ilr)}{r} \chi_p(\mathbf{x}) \tag{5.5.12}$$

where

$$f(\hat{\mathbf{r}}|\mathbf{k}n \to p) = -\frac{\mu}{2\pi\hbar^2} \int d\mathbf{s} \int d\mathbf{y} \exp(-il\hat{\mathbf{r}}\cdot\mathbf{s})\, \chi_p^*(\mathbf{y}) \times$$
$$\times V(\mathbf{s}, \mathbf{y})\, \Psi^+(\mathbf{s}, \mathbf{y}). \tag{5.5.13}$$

Equation (5.5.12) shows that Ψ^+ satisfies the asymptotic condition (4.1.24) and that the scattering amplitude is given by (5.5.13). We therefore have the important result that the Schwinger–Lippmann equation automatically gives the correct boundary conditions, including the fact that the sum over final states goes only over those which are energetically possible. In addition we see that the terms of the sum $\sum_p$ corresponding to energetically unobtainable states must decay exponentially as $r \to \infty$, which confirms the boundary condition imposed on the functions F_p introduced in Section 4.2 in the case of electron–hydrogen scattering.

The direction of the scattered particle is that of the unit vector $\hat{\mathbf{r}}$, and so the final wave number $\mathbf{l}$ must equal $l\hat{\mathbf{r}}$, while the final state Φ_β is given by

$$\Phi_\beta(\mathbf{r}, \mathbf{x}) = \exp(i\mathbf{l}\cdot\mathbf{r})\, \chi_p(\mathbf{x}). \tag{5.5.14}$$

We can therefore rewrite (5.5.13) as

$$\boxed{f(\Phi_\alpha \to \Phi_\beta) = -\frac{\mu}{2\pi\hbar^2} \langle \Phi_\beta | V | \Psi_\alpha{}^+ \rangle} \tag{5.5.15}$$

where we have attached the subscript α to Ψ^+ to emphasize that it is the Schwinger–Lippmann state corresponding to an unperturbed state Φ_α. The expression (5.5.15) is the generalization of the result (2.2.10) to the case of collisions of two composite particles.

When the collision energy $E \to +\infty$, an increasing number of states χ_p become energetically possible, and so the summation $\sum_p{}'$ becomes dominant. Further, the oscillation of the exponentials in the expression (5.5.6) for the Green's function suggests that the scattered wave

$$\int d\mathbf{s} \int d\mathbf{y}\; G_0^+(\mathbf{r}, \mathbf{x}; \mathbf{s}, \mathbf{y})\; V(\mathbf{s}, \mathbf{y})\; \Psi^+(\mathbf{s}, \mathbf{y})$$

tends to zero as $E \to \infty$ (see Section 2.3) and so

$$\Psi_\alpha^+(\mathbf{r}, \mathbf{x}) \underset{E\to\infty}{\sim} \Phi_\alpha(\mathbf{r}, \mathbf{x}). \tag{5.5.16}$$

This would also seem reasonable on physical grounds, for at high energies we should expect little scattering by the collision. We may therefore replace the exact expression (5.5.15) at high energies by the approximate expression

$$\boxed{f(\Phi_\alpha \to \Phi_\beta) \simeq -\frac{\mu}{2\pi\hbar^2}\langle \Phi_\beta | V | \Phi_\alpha \rangle.} \tag{5.5.17}$$

The Born series (2.4.9) for Ψ^+, and (2.4.14) for f, follow by exactly the same arguments as those of Section 2.4, if we put $\mathsf{G}_0 = (E - \mathsf{H}_0 + i\varepsilon)^{-1}$.

The reader may verify for himself that (5.4.13) is also an integral equation, this time for Ψ_α^-, whose asymptotic form is the sum of the initial state Φ_α and incoming waves.

If we substitute for $\Psi_\alpha^+ = \Psi^+$ from (5.4.15) into (5.5.15), the exact expression becomes

$$\boxed{f(\Phi_\alpha \to \Phi_\beta) = -\frac{\mu}{2\pi\hbar^2}\langle \Phi_\beta | V\Omega^+ | \Phi_\alpha \rangle.} \tag{5.5.18}$$

The *transition operator* T is defined by

$$\boxed{\mathsf{T} \equiv V\Omega^+,} \tag{5.5.19}$$

and it follows from (5.4.14) that

$$\mathsf{T} = V + V(E - \mathsf{H} + i\varepsilon)^{-1}V. \tag{5.5.20}$$

Thus (5.5.18) becomes

$$f(\Phi_\alpha \to \Phi_\beta) = -\frac{\mu}{2\pi\hbar^2}\langle\Phi_\beta|\mathsf{T}|\Phi_\alpha\rangle \tag{5.5.21}$$

showing that the scattering amplitude is proportional to the matrix element of the transition operator between the initial state Φ_α and final state Φ_β of the same energy, which accounts for its name.

5.6. Scattering of a particle by two centres of force

We shall now consider the problem of the scattering of a structureless particle A by two fixed centres P_1, P_2, which we suppose to be the sources of potentials V_1 and V_2 respectively. The situation is shown in Fig. 5.4. This will illustrate the use of formal scattering theory in the description of physical situations, derivation of approximations, and explanation of observed physical phenomena.

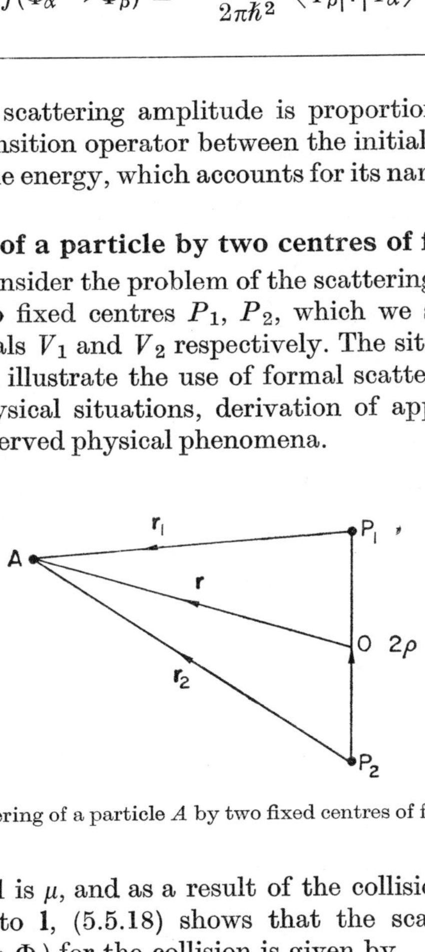

FIG. 5.4. Scattering of a particle A by two fixed centres of force P_1 and P_2.

If the mass of A is μ, and as a result of the collision its wave vector changes from $\mathbf{k}$ to $\mathbf{l}$, (5.5.18) shows that the scattering amplitude $f(\mathbf{k} \to \mathbf{l}) \equiv f(\Phi_\mathbf{k} \to \Phi_\mathbf{l})$ for the collision is given by

$$f(\mathbf{k} \to \mathbf{l}) = (-\mu/2\pi\hbar^2)\langle\Phi_\mathbf{l}|V\Omega^+|\Phi_\mathbf{k}\rangle. \tag{5.6.1}$$

where by (5.4.14)

$$\Omega^+ = 1+(E-\mathsf{H}+i\varepsilon)^{-1}\, V \tag{5.6.2}$$

and as usual $E = E_k = E_l$ is the energy of the particle. Since the potential V must be given by the expression

$$V = V_1 + V_2, \tag{5.6.3}$$

(5.6.2) can be rewritten

$$\Omega^+ = 1+(E-\mathsf{K}-V_1-V_2+i\varepsilon)^{-1}\,(V_1+V_2) \tag{5.6.4}$$

where K is the kinetic energy operator, while (5.6.1) can be rewritten

$$f(\mathbf{k}\to\mathbf{l}) = (-\mu/2\pi\hbar^2)\,\{\langle\Phi_{\mathbf{l}}|V_1\,\Omega^+|\Phi_{\mathbf{k}}\rangle+\langle\Phi_{\mathbf{l}}|V_2\,\Omega^+|\Phi_{\mathbf{k}}\rangle\}. \tag{5.6.5}$$

In these expressions $\Phi_{\mathbf{k}}(\mathbf{r}) = \exp(i\mathbf{k}\cdot\mathbf{r})$, $\Phi_{\mathbf{l}}(\mathbf{r}) = \exp(i\mathbf{l}\cdot\mathbf{r})$, where $\mathbf{r}$ is the position vector of A relative to the centre 0 of P_1P_2, which we regard as the scattering centre.

Now we can define Möller operators Ω_1^+ and Ω_2^+ by

$$\Omega_i^{\pm} \equiv 1+(E-\mathsf{K}-V_i\pm i\varepsilon)^{-1}\, V_i, \quad (i = 1,\, 2). \tag{5.6.6}$$

Ω_1^+ refers to scattering by V_1 alone, and Ω_2^+ refers to scattering by V_2 alone. Corresponding to these we can define transition operators T_1 and T_2 according to

$$\mathsf{T}_i = V_i\,\Omega_i^+, \quad (i = 1,\, 2). \tag{5.6.7}$$

We shall investigate the relationship between the transition operator T for the overall scattering and the transition operators T_1 and T_2 for scattering by the individual centres P_1 and P_2.

If we operate on (5.6.4) to the left with $E-\mathsf{K}-V_1-V_2+i\varepsilon$ and rearrange we have

$$(E-\mathsf{K}-V_1+i\varepsilon)\Omega^+ = (E-\mathsf{K}-V_1+i\varepsilon)+V_1+V_2\,\Omega^+. \tag{5.6.8}$$

Operation on (5.6.8) to the left with $(E-\mathsf{K}-V_1+i\varepsilon)^{-1}$ and use of (5.6.6) gives

$$\Omega^+ = \Omega_1^+ +(E-\mathsf{K}-V_1+i\varepsilon)^{-1}\, V_2\Omega^+. \tag{5.6.9}$$

Similarly we may show that

$$\Omega^+ = \Omega_2^+ +(E-\mathsf{K}-V_2+i\varepsilon)^{-1}\, V_1\Omega^+. \tag{5.6.10}$$

If we operate on (5.6.9) to the left with V_1, (5.6.10) to the left with V_2, add the results, and use the facts that $\mathsf{T} = V\Omega^+ = (V_1+V_2)\Omega^+$,

$\mathsf{T}_1 = V_1\Omega_1^+$, $\mathsf{T}_2 = V_2\Omega_2^+$, we see that

$$\mathsf{T} = \mathsf{T}_1+\mathsf{T}_2+V_1(E-\mathsf{K}-V_1+i\varepsilon)^{-1}V_2\Omega^+ + V_2(E-\mathsf{K}-V_2+i\varepsilon)^{-1}\,V_1\Omega^+. \tag{5.6.11}$$

Equation (5.6.11) expresses the relationship between the transition operator T for simultaneous scattering by both centres and the transition operators T_1 and T_2 for scattering by P_1 and P_2 alone. The transition matrix elements are therefore related by

$$\langle\Phi_\mathbf{l}|\mathsf{T}|\Phi_\mathbf{k}\rangle = \langle\Phi_\mathbf{l}|\mathsf{T}_1|\Phi_\mathbf{k}\rangle+\langle\Phi_\mathbf{l}|\mathsf{T}_2|\Phi_\mathbf{k}\rangle + \langle\Phi_\mathbf{l}|V_1(E-\mathsf{K}-V_1+i\varepsilon)^{-1}\,V_2\Omega^+|\Phi_\mathbf{k}\rangle + \langle\Phi_\mathbf{l}|V_2(E-\mathsf{K}-V_2+i\varepsilon)^{-1}\,V_1\Omega^+|\Phi_\mathbf{k}\rangle, \tag{5.6.12}$$

and we must now discover the physical significance of this expression.

Let $\Phi_\mathbf{k}^{(1)}$ be defined by

$$\Phi_\mathbf{k}^{(1)}(\mathbf{r}) = \exp\,(i\mathbf{k}\cdot\mathbf{r}_1) \tag{5.6.13}$$

where $\mathbf{r}_1$ is the position vector of A relative to P_1 (see Fig. 5.4). Since $\mathbf{r} = \boldsymbol{\rho}+\mathbf{r}_1$ where $2\boldsymbol{\rho} = \overrightarrow{P_2P_1}$, we have

$$\Phi_\mathbf{k} = \exp\,(i\mathbf{k}\cdot\boldsymbol{\rho})\;\Phi_\mathbf{k}^{(1)} \tag{5.6.14}$$

and the same is obviously true with $\mathbf{k}$ replaced by $\mathbf{l}$. It therefore follows that

$$\langle\Phi_\mathbf{l}|\mathsf{T}_1|\Phi_\mathbf{k}\rangle = \exp\,(i\mathbf{q}\cdot\boldsymbol{\rho})\;\langle\Phi_\mathbf{l}^{(1)}|\mathsf{T}_1|\Phi_\mathbf{k}^{(1)}\rangle \tag{5.6.15}$$

where $\mathbf{q} = \mathbf{k}-\mathbf{l}$. But

$$\langle\Phi_\mathbf{l}^{(1)}|\mathsf{T}_1|\Phi_\mathbf{k}^{(1)}\rangle = \langle\Phi_\mathbf{l}^{(1)}|\,V_1\Omega_1^+|\Phi_\mathbf{k}^{(1)}\rangle \tag{5.6.16}$$

and since V_1 is centred on P_1, the scattering amplitude produced by it is

$$f_1(\mathbf{k}\to\mathbf{l}) = (-\mu/2\pi\hbar^2)\;\langle\Phi_\mathbf{l}^{(1)}|\mathsf{T}_1|\Phi_\mathbf{k}^{(1)}\rangle. \tag{5.6.17}$$

Hence from (5.6.15) we have

$$(-\mu/2\pi\hbar^2)\;\langle\Phi_\mathbf{l}|\mathsf{T}_1|\Phi_\mathbf{k}\rangle = \exp\,(i\mathbf{q}\cdot\boldsymbol{\rho})\,f_1(\mathbf{k}\to\mathbf{l}). \tag{5.6.18}$$

Similarly, we find

$$(-\mu/2\pi\hbar^2)\;\langle\Phi_\mathbf{l}|\mathsf{T}_2|\Phi_\mathbf{k}\rangle = \exp\,(-i\mathbf{q}\cdot\boldsymbol{\rho})\,f_2(\mathbf{k}\to\mathbf{l}) \tag{5.6.19}$$

where f_2 is the amplitude for scattering by the centre P_2.

From (5.6.18) and (5.6.19) we see that multiplication of (5.6.12) by $(-\mu/2\pi\hbar^2)$ gives us, on use of (5.6.1),

$$\begin{aligned} f(\mathbf{k}\to\mathbf{l}) = & \exp(i\mathbf{q}\cdot\boldsymbol{\rho})\,f_1(\mathbf{k}\to\mathbf{l})+\exp(-i\mathbf{q}\cdot\boldsymbol{\rho})\,f_2(\mathbf{k}\to\mathbf{l}) \\ & +(-\mu/2\pi\hbar^2)\,\langle\Phi_{\mathbf{l}}|V_1(E-\mathsf{K}-V_1+i\varepsilon)^{-1}\,V_2\Omega^+|\Phi_{\mathbf{k}}\rangle \\ & +(-\mu/2\pi\hbar^2)\,\langle\Phi_{\mathbf{l}}|V_2(E-\mathsf{K}-V_2+i\varepsilon)^{-1}\,V_1\Omega^+|\Phi_{\mathbf{k}}\rangle. \end{aligned} \tag{5.6.20}$$

This shows that the scattering amplitude is the sum of the scattering amplitudes from P_1 and P_2 multiplied by certain phase factors depending on the distance between P_1 and P_2, together with two correction terms. What is the significance of these correction terms?

We have from (5.6.14) with $\mathbf{k} = \mathbf{l}$ that, since $\Omega^+\Phi_{\mathbf{k}} = \Psi^+_{\mathbf{k}}$,

$$\begin{aligned} & \langle\Phi_{\mathbf{l}}|V_1(E-\mathsf{K}-V_1+i\varepsilon)^{-1}\,V_2\Omega^+|\Phi_{\mathbf{k}}\rangle \\ & \quad = \exp(-i\mathbf{l}.\boldsymbol{\rho})\,\langle(E-\mathsf{K}-V_1-i\varepsilon)^{-1}\,V_1\Phi^{(1)}_{\mathbf{l}}|V_2|\Psi^+_{\mathbf{k}}\rangle, \end{aligned} \tag{5.6.21}$$

and by (5.6.6) this may be rewritten

$$\begin{aligned} & \langle\Phi_{\mathbf{l}}|V_1(E-\mathsf{K}-V_1+i\varepsilon)^{-1}\,V_2\Omega^+|\Phi_{\mathbf{k}}\rangle \\ & \quad = \exp(-i\mathbf{l}\cdot\boldsymbol{\rho})\,\langle(\Omega^-_1-1)\Phi^{(1)}_{\mathbf{l}}|V_2|\Psi^+_{\mathbf{k}}\rangle. \end{aligned} \tag{5.6.22}$$

In order to determine the order of magnitude of this term we note that, by definition,

$$\Psi^+_{-\mathbf{k}} = \Phi_{-\mathbf{k}}+(E-\mathsf{H}+i\varepsilon)^{-1}\,V\Phi_{-\mathbf{k}};$$

if we take the complex conjugate of this and note that $\Phi^*_{-\mathbf{k}} = \Phi_{\mathbf{k}}$ we see that

$$\Psi^{+*}_{-\mathbf{k}} = \Phi_{\mathbf{k}}+(E-\mathsf{H}-i\varepsilon)^{-1}\,V\Phi_{-\mathbf{k}}.$$

The right-hand side of this is simply $\Psi^-_{\mathbf{k}}$, and so we obtain the result

$$\boxed{\Psi^-_{\mathbf{k}} = \Psi^{+*}_{-\mathbf{k}}.} \tag{5.6.23}$$

Since $\Omega^\pm\,\Phi_{\mathbf{k}} = \Psi^\pm_{\mathbf{k}}$, (5.6.23) shows that

$$(\Omega^- -1)\Phi_{\mathbf{k}} = (\Psi^+_{\mathbf{k}}-\Phi_{-\mathbf{k}})^* = \Psi^{s*}_{-\mathbf{k}} \tag{5.6.24}$$

where $\Psi^s_{-\mathbf{k}}$ is the scattered wave corresponding to incident momentum $-\hbar\mathbf{k}$. We can now use (5.6.24) to rewrite (5.6.22) as

$$\begin{aligned} & \langle\Phi_{\mathbf{l}}|V_1(E-\mathsf{K}-V_1+i\varepsilon)^{-1}\,V_2\Omega^+|\Phi_{\mathbf{k}}\rangle \\ & \quad = \exp(-i\mathbf{l}\cdot\boldsymbol{\rho})\,\langle\Psi^{(1)s*}_{-\mathbf{l}}|V_2|\Psi^+_{\mathbf{k}}\rangle \end{aligned} \tag{5.6.25}$$

where $\Psi^{(1)s}_{-\mathbf{l}}$ is the wave scattered by V_1 with P_1 regarded as the centre of force. Hence if this is small in the region of P_2, where V_2 is big, we would expect this term to be small. The order of magnitude of this wave at P_2 is $|\bar{f}_1|/(2\rho)$ where $|\bar{f}_1|$ is the average value of the amplitude for scattering by the centre P_1, and this is roughly $(\sigma_1/4\pi)^{\frac{1}{2}}/(2\rho)$, where σ_1 is the total cross-section for scattering by P_1. Hence if

$$(\sigma_1/4\pi)^{\frac{1}{2}}/(2\rho) \ll 1 \tag{5.6.26}$$

we expect the first correction term to be small compared with the previous terms on the right-hand side of (5.6.20).

In a similar way, if

$$(\sigma_2/4\pi)^{\frac{1}{2}}/(2\rho) \ll 1 \tag{5.6.27}$$

where σ_2 is the cross-section for scattering by P_2, we expect the fourth term on the right-hand side of (5.6.20) to be small; (5.6.20) therefore takes the final form

$$f(\mathbf{k} \to \mathbf{l}) \simeq \exp(i\mathbf{q} \cdot \boldsymbol{\rho})\, f_1(\mathbf{k} \to \mathbf{l}) + \exp(-i\mathbf{q} \cdot \boldsymbol{\rho})\, f_2(\mathbf{k} \to \mathbf{l}), \tag{5.6.28}$$

which is an approximate expression for the scattering amplitude when "multiple scattering effects" are neglected. If we recall that the differential cross-section is the square of the modulus of the scattering amplitude, (5.6.28) gives us

$$\sigma(\mathbf{k} \to \mathbf{l}) \simeq \sigma_1(\mathbf{k} \to \mathbf{l}) + \sigma_2(\mathbf{k} \to \mathbf{l}) + 2\mathrm{Re}[\exp(2i\mathbf{q} \cdot \boldsymbol{\rho})\, f_1(\mathbf{k} \to \mathbf{l})\, f_2^*(\mathbf{k} \to \mathbf{l})] \tag{5.6.29}$$

where $\sigma_1(\mathbf{k} \to \mathbf{l})$ and $\sigma_2(\mathbf{k} \to \mathbf{l})$ are the differential cross-sections for scattering by the centres P_1 and P_2 alone.

The classical result for scattering from two centres when multiple scattering effects are ignored is the same as (5.6.29) apart from the third term. It is this which gives rise to interference patterns in the scattering of electrons—a result essentially wave-mechanical in its nature.

REFERENCES

DIRAC, P. A. M. (1958) *The Principles of Quantum Mechanics,* Oxford University Press.

LIPPMANN, B. A. and SCHWINGER, J. (1950) *Phys. Rev.* **79**, 469.

NEWTON, R. G. (1966) *Scattering Theory of Waves and Particles,* McGraw-Hill.

CHAPTER 6

SCATTERING OF A WAVE PACKET BY A CENTRE OF FORCE

6.1. Solution of Schrödinger's time-dependent equation for a free particle; wave packets

We have so far considered the collision problem from a time-independent viewpoint. To solve the problem of scattering of a single particle by a centre of force O we looked for a solution Ψ of the time-independent Schrödinger equation (1.2.10) which was regular at the origin and satisfied the boundary condition (1.2.11). The differential cross-section was then calculated from (1.2.9).

This approach suffers from two important disadvantages. In the first place a collision is a dynamic process which takes place in time, for a particle comes in from a source such as a linear accelerator or cyclotron, enters the target chamber, interacts with a target particle, and then leaves the chamber and is collected by a detector. Such a process cannot be properly described unless the state of the particle is considered to change with time.

The second disadvantage is mathematical. Quantum mechanical states are assumed to be elements of Hilbert space (Volume 1, Section 3.4), and must therefore be normalizable. Hence if ψ is any wave function the quantity

$$\|\psi\| = \{\int |\psi(\mathbf{r})|^2 \, d\mathbf{r}\}^{\frac{1}{2}}, \tag{6.1.1}$$

where the integral is taken over all space, must be finite. Since $|\psi(\mathbf{r})|^2 \, d\mathbf{r}$ is interpreted as the probability of finding the particle in the volume element $d\mathbf{r}$ at point $\mathbf{r}$, the value of the integral is the probability of finding the particle somewhere, and is therefore unity. In the case of the scattering wave function Ψ satisfying the boundary condition (1.2.11) the integral in (6.1.1) clearly does not exist. Many mathematical properties such as Hermiticity of various operators, completeness of sets of eigenfunctions, and others, may not hold for functions which are not square integrable; that is to say, for functions for which the integral in (6.1.1) diverges or does not exist.

We see therefore that both mathematically and physically a much more correct procedure for the description of collision processes in quantum mechanics is to use *wave packets*. A wave packet, which we shall denote by $\bar{\psi}$, is a function of the coordinates $\mathbf{q}$ of the system and of the time t which satisfies the time-dependent Schrödinger equation

$$i\hbar \frac{\partial \bar{\psi}}{\partial t} = \mathsf{H}\bar{\psi} \tag{6.1.2}$$

and the normalization condition

$$\int |\bar{\psi}(\mathbf{q}, t)|^2 \, d\mathbf{q} = 1. \tag{6.1.3}$$

We shall show later that if $\bar{\psi}$ satisfies the condition (6.1.3) at any one instant of time it satisfies it at all other times. If $\bar{\psi}$ is a bound state it has the form

$$\bar{\psi}(\mathbf{q}, t) = \psi_n(\mathbf{q}) \exp(-iE_n t/\hbar) \tag{6.1.4}$$

where $\mathsf{H}\psi_n = E_n\psi_n$, E_n being the energy of the state and $\|\psi_n\| = 1$. This satisfies (6.1.3) at all times. Scattering states we shall see are not states of exact energy but are wave packets which move in space. As the wave packet travels through space the region in which the particle is likely to be found changes with time.

Throughout this chapter we shall develop this approach in the simplest case, which we first considered in Chapter 1 from a time-independent point of view, of the scattering of a single structureless particle by a fixed centre O. The scattering potential $V = V(\mathbf{r})$ need not necessarily be spherically symmetric. The Hamiltonian is therefore given by

$$\mathsf{H} = \mathsf{H}_0 + V = (-\hbar^2/2\mu)\nabla^2 + V. \tag{6.1.5}$$

In this section we shall obtain a general expression for any wave packet $\bar{\varphi} = \bar{\varphi}(\mathbf{r}, t)$ describing the motion of a free particle when $V = 0$, and consider some of its properties. In the next section we shall investigate the nature of wave packets produced in actual scattering experiments, and subsequent sections will follow their evolution in time when they begin to overlap the region of interaction in which $V \neq 0$ (assumed finite).

When $V = 0$, $\bar{\psi} = \bar{\varphi}$, so that (6.1.2) and (6.1.5) give

$$i\hbar \frac{\partial \bar{\varphi}}{\partial t} = -\frac{\hbar^2}{2\mu} \nabla^2 \bar{\varphi}. \tag{6.1.6}$$

$\bar{\varphi}$ must be the Fourier transform of some function $C(\mathbf{k}, t)$ so that

$$\bar{\varphi}(\mathbf{r}, t) = (2\pi)^{-3/2} \int d\mathbf{k} \, C(\mathbf{k}, t) \exp(i\mathbf{k}\cdot\mathbf{r}). \tag{6.1.7}$$

If we substitute for $\bar{\varphi}$ from (6.1.7) into (6.1.6) we have, remembering that $E_k \equiv (\hbar^2 k^2/2\mu)$,

$$\int d\mathbf{k}\, i\hbar \frac{\partial C(\mathbf{k}, t)}{\partial t} \exp(i\mathbf{k}\cdot\mathbf{r}) = \int d\mathbf{k}\, E_k\, C(\mathbf{k}, t) \exp(i\mathbf{k}\cdot\mathbf{r}). \quad (6.1.8)$$

Let us multiply (6.1.8) by $\exp(-i\mathbf{l}\cdot\mathbf{r})$ and integrate over all values of $\mathbf{r}$ using the relation (Vol. 1, pp. 112–13)

$$\int d\mathbf{r} \exp[i(\mathbf{k}-\mathbf{l})\cdot\mathbf{r}] = (2\pi)^3\, \delta(\mathbf{k}-\mathbf{l}). \quad (6.1.9)$$

This gives the equation

$$i\hbar \frac{\partial C(\mathbf{l}, t)}{\partial t} = E_l\, C(\mathbf{l}, t). \quad (6.1.10)$$

For fixed $\mathbf{l}$ we solve this differential equation by the elementary method of separating variables to obtain

$$\log[C(\mathbf{l}, t)] = (-iE_l\, t/\hbar) + \text{const};$$

this is true for all $\mathbf{l}$, and therefore in particular when $\mathbf{l} = \mathbf{k}$, and so we obtain

$$C(\mathbf{k}, t) = C(\mathbf{k}) \exp(-iE_k\, t/\hbar) \quad (6.1.11)$$

where $C(\mathbf{k}) \equiv C(\mathbf{k}, 0)$ is some function of $\mathbf{k}$. It follows from (6.1.7) that

$$\bar{\varphi}(\mathbf{r}, t) = (2\pi)^{-3/2} \int d\mathbf{k}\, C(\mathbf{k}) \exp(i\mathbf{k}\cdot\mathbf{r} - iE_k\, t/\hbar) \quad (6.1.12)$$

which is the solution of the time-dependent problem for a free particle.

The result (6.1.12) may also be written in the form

$$\bar{\varphi}(\mathbf{r}, t) = \int d\mathbf{k}\, C(\mathbf{k})\, \varphi_{\mathbf{k}}(\mathbf{r}) \exp(-iE_k\, t/\hbar). \quad (6.1.13)$$

We note that (6.1.7) is an expansion of $\bar{\varphi}$ in terms of the δ-function normalized eigenfunctions $(2\pi)^{-3/2} \exp(i\mathbf{k}\cdot\mathbf{r})$ of the momentum operator $\mathsf{p} \equiv -i\hbar\,\nabla$, and it follows that the probability $P(\mathbf{k})\, d\mathbf{k}$ of finding the particle with wave vector $\mathbf{k}$ lying in the element of volume $(\mathbf{k}, d\mathbf{k})$ of $\mathbf{k}$-space is

$$P(\mathbf{k}) = |C(\mathbf{k}, t)|^2. \quad (6.1.14)$$

From (6.1.11) we see that

$$P(\mathbf{k}) = |C(\mathbf{k})|^2, \quad (6.1.15)$$

and so $P(\mathbf{k})$ is constant in time, as is to be expected from the fact that p commutes with the free particle Hamiltonian $\mathsf{H}_0 = \mathsf{p}^2/2\mu$. These results follow at once from the basic principles (Volume 1, Section 4.7).

The probability of the particle being observed to have some momentum $\hbar\mathbf{k}$ must be unity, and so (6.1.15) implies that

$$\int d\mathbf{k}|C(\mathbf{k})|^2 = 1. \tag{6.1.16}$$

It follows from (6.1.12), (6.1.9) and (6.1.16) that

$$\begin{aligned}
&\int d\mathbf{r}|\bar{\varphi}(\mathbf{r}, t)|^2\, d\mathbf{r} \\
&\quad = \int d\mathbf{r}\, \{(2\pi)^{-3/2} \int d\mathbf{k}\, C^*(\mathbf{k}) \exp(-i\mathbf{k}\cdot\mathbf{r} + iE_k\, t/\hbar) \times \\
&\qquad \times (2\pi)^{-3/2} \int d\mathbf{l}\, C(\mathbf{l}) \exp(i\mathbf{l}\cdot\mathbf{r} - iE_l\, t/\hbar)\} \\
&\quad = \int d\mathbf{k}|C(\mathbf{k})|^2 = 1,
\end{aligned} \tag{6.1.17}$$

confirming (6.1.3) in this special case.

Let $\mathbf{k}$ have components k_x, k_y, k_z relative to Cartesian axes Ox, Oy, Oz; then (6.1.12) can be written more fully as

$$\begin{aligned}
\bar{\varphi}(x, y, z, t) = (2\pi)^{-3/2} \int_{-\infty}^{+\infty} dk_x \int_{-\infty}^{+\infty} dk_y \int_{-\infty}^{+\infty} dk_z\, C(k_x, k_y, k_z) \times \\
\times \exp(ik_x x + ik_y y + ik_z z - iE_k\, t/\hbar)
\end{aligned} \tag{6.1.18}$$

where

$$E_k = \hbar^2(k_x{}^2 + k_y{}^2 + k_z{}^2)/2\mu. \tag{6.1.19}$$

Now consider what happens to the integral on the right-hand side of (6.1.18) as $t \to \pm\infty$. The rate of change of the argument of the exponential with respect to k_z, for example, is

$$(\partial/\partial k_z)\, \{k_x x + k_y y + k_z z - E_k\, t/\hbar\} = z - \hbar k_z\, t/\mu. \tag{6.1.20}$$

As $t \to \pm\infty$ this tends to $\mp\infty$, which means that the oscillation of the exponential as k_z varies becomes infinitely fast, and so as in Section 2.3 we see that the integral must tend to zero. Thus we obtain the result

$$\bar{\varphi}(\mathbf{r}, t) \to 0 \quad \text{as} \quad t \to \pm\infty. \tag{6.1.21}$$

In other words, the probability of finding the particle in any finite region tends to zero as $t \to \pm\infty$.

This last result has a simple physical interpretation. There will always be an uncertainty Δv in the speed of the particle, and consequently a growing uncertainty $\Delta v t$ in its position. Thus the region in which the particle may be found grows indefinitely large, and the probability of it being found in any particular region tends to zero.

6.2. Experimental wave packets

In an actual scattering experiment a particle is prepared in its initial state by an emitting apparatus. This usually consists of a source of incident particles, for example a furnace in which ions are prepared, a cyclotron or linear accelerator to give the particles the required energy, and a collimator to ensure a parallel beam. In addition, the experimentalist usually attempts to produce a uniform beam. These facts enable us to draw some conclusions about the nature of a wave packet actually produced in the laboratory.

We have already seen in the previous section that any free wave packet must be given by an expression of the form (6.1.13) where $|C(\mathbf{k})|^2\, d\mathbf{k}$ is the probability of finding the particle with a wave vector in the volume element $(\mathbf{k}, d\mathbf{k})$ of $\mathbf{k}$-space. As in Chapter 1 we take Cartesian axes with Oz parallel to the axis of the incident beam. An experiment is usually designed to produce bombarding particles with energy E_0 and velocity v_0 parallel to Oz, so that $E_0 = \frac{1}{2}\mu v_0{}^2$. This corresponds to a momentum μv_0 parallel to the z-axis, giving rise to a wave vector $\mathbf{k}_0 = \mu\mathbf{v}_0/\hbar$ where $\mathbf{k}_0$, $\mathbf{v}_0$ have components $(0, 0, k_0)$, $(0, 0, v_0)$ relative to $Oxyz$.

Now in practice there will be neither perfect collimation nor perfect energy resolution. In the x-direction the mean velocity is zero, but there will be a margin of error Δv_x in the velocity, corresponding to a maximum likely angular deviation of $\Delta v_x/v_0$. If collimation is good, $\Delta v_x/v_0 \ll 1$. The error $\mu\Delta v_x$ in momentum parallel to Ox gives rise to a variation Δk_x in the wave vector, where $\hbar\Delta k_x = \mu\Delta v_x$. Since $C(\mathbf{k})$ is the probability amplitude for the wave vector $\mathbf{k}$, we know that this must vanish if $k_x > \Delta k_x$, for there is no probability of observing an angular deviation greater than $\Delta v_x/v_0 = \Delta k_x/k_0$. Furthermore, if the collimation is good the maximum angular deviation of the beam is small, and so $\Delta k_x \ll k_0$. Similarly in the y-direction there is a spread of momentum $\hbar\Delta k_y$, so that $C(\mathbf{k})$ vanishes if $|k_y| > \Delta k_y$, and $\Delta k_y \ll k_0$.

The mean energy is $E_0 = \hbar^2 k_0{}^2/2\mu$, and so there is a spread of energy of amount

$$\Delta E \equiv \Delta(\hbar^2 k^2/2\mu)_{(\mathbf{k}=\mathbf{k}_0)} \simeq \hbar^2 k_0\, \Delta k_z/\mu, \qquad (6.2.1)$$

where $\hbar\Delta k_z$ is the spread of momentum in the z-direction so that $C(\mathbf{k}) = 0$ if $|k_z - k_0| > \Delta k_z$. Since $\Delta E/E_0 = {}_2\Delta k_z/k_0$, we see that if the energy resolution is good, then $\Delta k_z \ll k_0$.

We now investigate the implications on the nature of the spatial wave functions of these properties of $C(\mathbf{k})$; such implications apply to a wave packet in the laboratory. To do this we note that (6.1.18) can be written

$$\bar{\varphi}(x, y, z, t) = (2\pi)^{-3/2} \int_{-\infty}^{+\infty} dk_x \int_{-\infty}^{+\infty} dk_y \int_{-\infty}^{+\infty} dk_z |C(\mathbf{k})| \times$$
$$\times \exp\,[ik_x x + ik_y y + ik_z z - iE_k\, t/\hbar + i \arg C(\mathbf{k})] \qquad (6.2.2)$$

where E_k is given by (6.1.19). The integrand of (6.2.2) consists of a non-negative function $|C(\mathbf{k})|$ of $\mathbf{k}$, which has a maximum when $\mathbf{k} = \mathbf{k}_0$ and diminishes smoothly to zero as $|k_x|$, $|k_y|$, $|k_z - k_0|$ increase steadily from zero to Δk_x, Δk_y, Δk_z respectively, multiplied by an exponential with imaginary exponent. The magnitude of $\bar{\varphi}$ will depend upon the oscillation of the exponential as k_x, k_y, k_z vary, and hence upon the values of the partial derivatives of the argument of the exponential with respect to k_x, k_y, and k_z.

Now the partial derivative of the argument with respect to k_z is

$$(\partial/\partial k_z)\,[k_x x + k_y y + k_z z - E_k\, t/\hbar + \arg C(\mathbf{k})]$$
$$= z - \hbar k_z\, t/\mu + (\partial/\partial k_z) \arg C(\mathbf{k}). \qquad (6.2.3)$$

We need only consider values of k_z lying between $k_0 - \Delta k_z$ and $k_0 + \Delta k_z$, for $C(\mathbf{k})$ vanishes outside this range. Let us suppose in the first place that $t < 0$; then the greatest value of the right-hand side of (6.2.3) is

$$\text{Max}\ (\partial/\partial k_z)\,[\mathbf{k}\cdot\mathbf{r} - E_k\, t/\hbar + \arg C(\mathbf{k})] = z - \hbar(k_0 + \Delta k_z)\, t/\mu + M_z \qquad (6.2.4)$$

where M_z is the maximum value of $(\partial/\partial k_z) \arg C(\mathbf{k})$ for values of $\mathbf{k}$ for which $|C(\mathbf{k})| \neq 0$. The minimum value is

$$\text{Min}\ (\partial/\partial k_z)\,[\mathbf{k}\cdot\mathbf{r} - E_k\, t/\hbar + \arg C(\mathbf{k})] = z - \hbar(k_0 - \Delta k_z)\, t/\mu + m_z \qquad (6.2.5)$$

where m_z is the corresponding minimum value of $(\partial/\partial k_z) \arg C(\mathbf{k})$.

Let n be a large, positive integer. If the rate of change of the argument of the exponential in (6.2.2) is greater than $+n\pi/\Delta k_z$ for all values of $\mathbf{k}$ for which $|C(\mathbf{k})| \neq 0$, the argument of the exponential will increase by at least $(2\Delta k_z) \times (n\pi/\Delta k_z) = 2n\pi$ as k_z increases from $k_0 - \Delta k_z$ to $k_0 + \Delta k_z$, which is the effective range of integration. Thus the exponential will oscillate at least n times during the integration, and since $|C(\mathbf{k})|$ is smooth this means that $\bar{\varphi}$ will be negligible for large values of n, and small even for $n = 1$ (cf. Fig. 2.1). The rate of change of the argument will be greater than $n\pi/\Delta k_z$ if the minimum value given by (6.2.5) is greater than $n\pi/\Delta k_z$, and so

$$z - \hbar(k_0 - \Delta k_z)t/\mu + m_z > n\pi/\Delta k_z \quad \text{implies} \quad \bar{\varphi}(\mathbf{r}, t) = 0. \tag{6.2.6}$$

It obviously follows from this that

$$z > \hbar(k_0 - \Delta k_z)t/\mu - m_z + n\pi/\Delta k_z \quad \text{implies} \quad \bar{\varphi}(r, t) = 0. \tag{6.2.7}$$

In a similar way we can see that if the maximum value of the rate of change is less than $-n\pi/\Delta k_z$, $\bar{\varphi}(\mathbf{r}, t)$ must be negligibly small, so that

$$z < \hbar(k_0 + \Delta k_z)t/\mu - M_z - n\pi/\Delta k_z \quad \text{implies} \quad \bar{\varphi}(\mathbf{r}, t) = 0. \tag{6.2.8}$$

The results (6.2.7) and (6.2.8) show that, for negative times, the particle must be confined to the region

$$\hbar(k_0 + \Delta k_z)t/\mu - M_z - n\pi/\Delta k_z < z < \hbar(k_0 - \Delta k_z)t/\mu - m_z + n\pi/\Delta k_z. \tag{6.2.9}$$

Since $\hbar k_0 = \mu v_0$, $\hbar\Delta k_z = \mu\Delta v_z$, (6.2.9) can be written

$$(v_0 + \Delta v_z)t - M_z - n\pi/\Delta k_z < z < (v_0 - \Delta v_z)t - m_z + n\pi/\Delta k_z. \tag{6.2.10}$$

Hence the particle is confined between the two planes

$$z = (v_0 + \Delta v_z)t - M_z - n\pi/\Delta k_z = Z_1 \text{ say,}$$

$$z = (v_0 - \Delta v_z)t - m_z + n\pi/\Delta k_z = Z_2 \text{ say,}$$

shown in Fig. 6.1. Both planes move to the right; the left-hand one with speed $v_0 + \Delta v_z$, the right-hand one with speed $v_0 - \Delta v_z$.

The plane midway between the planes $z = Z_1$ and $z = Z_2$ is $z = Z$ where $Z = \frac{1}{2}(Z_1 + Z_2)$, so that

$$Z \equiv v_0 t - \tfrac{1}{2}(m_z + M_z). \tag{6.2.11}$$

Hence the central plane moves with velocity v_0 to the right.

The distance between the planes, which is the length of the wave packet in the z-direction, is given by $d_z \equiv Z_2 - Z_1$, and so

$$d_z = -2\Delta v_z t + (M_z - m_z) + 2n\pi/\Delta k_z. \tag{6.2.12}$$

Now $\hbar\Delta k_z = \Delta p_z$ where p_z is the z-component of momentum. If $p_0 = \hbar k_0$ is the mean momentum, the range of momenta likely to be observed is $p_0 - \Delta p_z < p_z < p_0 + \Delta p_z$. Further, $2\pi\hbar = h$, and so (6.2.12) can be rewritten as

$$d_z = -2\Delta v_z t + (M_z - m_z) + nh/\Delta p_z, \quad (t < 0). \tag{6.2.13}$$

Let us compare this result with the classical case. Suppose at time $t = 0$ we knew that the particle P was somewhere between the planes $z = a$ and $z = b$ $(a < b)$, and that its velocity parallel to Oz was somewhere between $v_0 - \Delta v_z$ and $v_0 + \Delta v_z$. Then at some other time $t < 0$ the smallest value of z that the particle could have had must be $(v_0 + \Delta v_z)t + a$, corresponding to P being at $z = a$ at time $t = 0$ (the farthest left possible) and with z-component of velocity $v_0 + \Delta v_z$ (the greatest speed possible). Similarly at time t the farthest right it could have been is $z = (v_0 - \Delta v_z)t + b$, corresponding to being farthest right

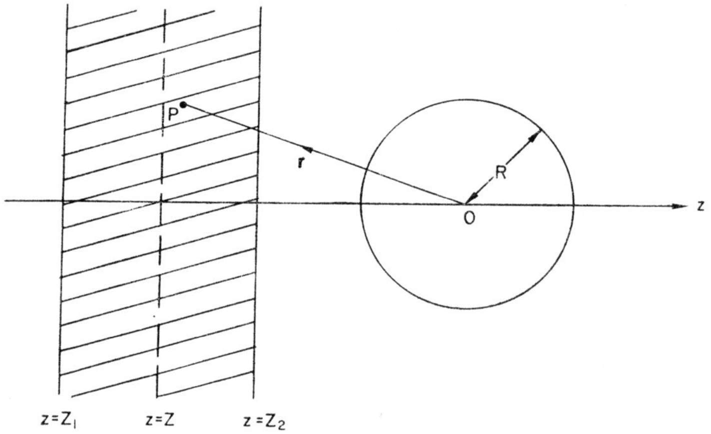

FIG. 6.1. Shape of free particle wave packet for $t < 0$; $Z_1 = (v_0 + \Delta v_z) t - M_z - n\pi/\Delta k_z$, $Z_2 = (v_0 - \Delta v_z)t - m_z + n\pi/\Delta k_z$, $Z = v_0 t - \frac{1}{2}(m_z + M_z)$.

at time $t = 0$ and minimum speed. Thus at time t (< 0) P must have been in the region

$$(v_0 + \Delta v_z)t + a < z < (v_0 - \Delta v_z)t + b, \tag{6.2.14}$$

the classical result corresponding to (6.2.10). The classical distance d_z^c between these planes is therefore given by

$$d_z^c = -2\Delta v_z t + (b - a) \tag{6.2.15}$$

which corresponds to (6.2.13). There is, however, an important difference. Classically there is nothing to prevent us from determining the position of P at time $t = 0$ with absolute precision, so that $b = a$, and in this case $d_z^c = -2\Delta v_z t = 0$ when $t = 0$. Quantum mechanically we can choose the function $C(\mathbf{k})$ so that $(\partial/\partial k_z)\arg C(\mathbf{k})$ is constant, when $M_z = m_z$ but Δv_z is unaffected. We then have from (6.2.13) that $d_z = -2\Delta v_z t + nh/\Delta p_z$, and this takes its minimum value $nh/\Delta p_z$ when $t = 0$. Since $n \geqq 1$, it follows that the z-width is always greater than or of the order of $h/\Delta p_z$, a residual uncertainty which is essentially quantum mechanical. It is, in fact, a consequence of the position-momentum uncertainty principle (Volume 1, Section 4.5).

Precisely similar arguments must apply to motion parallel to Ox and Oy with v_0 replaced by zero, and so the x- and y-coordinates of P must lie in intervals whose mid-points are

$$X = -\tfrac{1}{2}(m_x + M_x), \quad Y = -\tfrac{1}{2}(m_y + M_y), \tag{6.2.16}$$

and whose lengths are

$$\begin{aligned} d_x &= -2\Delta v_x t + (M_x - m_x) + nh/\Delta p_x, \\ d_y &= -2\Delta v_y t + (M_y - m_y) + nh/\Delta p_y. \end{aligned} \tag{6.2.17}$$

In (6.2.16) and (6.2.17) Δp_x is the uncertainty in momentum parallel to Ox, etc.

If we put $\mathbf{R}$ equal to the vector (X, Y, Z) and $\mathbf{m} = (m_x, m_y, m_z)$, $\mathbf{M} = (M_x, M_y, M_z)$, $\mathbf{v}_0 = (0, 0, v_0)$, so that $\mathbf{v}_0$ is the mean velocity, we see that (6.2.11) and (6.2.16) may be summarized by the formula

$$\mathbf{R} = \mathbf{v}_0 t - \tfrac{1}{2}(\mathbf{m} + \mathbf{M}), \tag{6.2.18}$$

while (6.2.13) and (6.2.17) may be summarized by

$$d_i = -2\Delta v_i t + (M_i - m_i) + nh/\Delta p_i, \quad (i = x, y, z). \tag{6.2.19}$$

The region to which the particle is confined is therefore a rectangular parallelopiped whose sides are of length d_x, d_y, d_z, and whose centre moves with velocity $\mathbf{v}_0$ and passes through the point $\mathbf{r} = -\tfrac{1}{2}(\mathbf{m} + \mathbf{M})$ at time $t = 0$. The rate of contraction in the ith direction is Δv_i, and since $\Delta v_i \ll v_0$ for $i = x, y, z$, the contraction of the wave packet is negligible compared with the motion of its centre. The nature of the wave packet when $t > 0$ may be similarly determined. It is not difficult to show that formula (6.2.18) is still valid, but that t in (6.2.19) must be replaced by $-t$.

The reader may wonder at our conclusion that there is a contraction for negative times, rather than an expansion. What we have proved, however, is that the wave packet which emerges from the emitting apparatus is confined to a rectangular parallelopiped which contracts and then expands, given the properties of $|C(\mathbf{k})|$ which are imposed by the nature of the collimating process. In fact knowledge of $|C(\mathbf{k})|$ does not give us complete knowledge of $C(\mathbf{k})$, for its argument is unspecified.

For this reason the quantity $\mathbf{M}_i - \mathbf{m}_i$ is unknown. If the system is prepared at some time t_0 in such a way that we have complete knowledge of the state, and hence of $C(\mathbf{k})$, the wave function develops subsequently according to the Schrödinger time-dependent equation. If the reader follows through the above argument the only difference he will find is that the time t is replaced by $t - t_0$. It therefore follows that the packet expands for $t > t_0$; that is, for times subsequent to the preparation with complete knowledge. We cannot talk about the development of the state for times previous to t_0, for this would then involve us in the preparation process when the Schrödinger equation is no longer applicable.

In conclusion we can say that the wave packet which emerges from the source is confined to a rectangular box whose centre moves with the mean velocity v_0 parallel to Oz and whose rates of expansion and contraction are negligible compared with the velocity of the centre. Initially the parallelopiped will lie entirely outside the sphere of radius R and centre O in which the interaction potential V is non-vanishing (see Fig. 6.1), and it will be possible to assume that the wave packet moves freely. At some time after the emission of the particle the rectangular parallelopiped will begin to overlap the region of interaction, and it will then be necessary to study the development of the wave packet under the full Hamiltonian $\mathsf{H} = \mathsf{H}_0 + V$. The object of the remainder of this chapter is to study the nature of this development. First, however, it is necessary to develop the mathematical methods for dealing with the time-evolution of wave functions.

6.3. Unitary operators

In this section we shall digress in order to recall‡ the properties of operators in "Hilbert space" which are fundamental to quantum theory.

The *inverse* A^{-1} of an operator A possesses the properties

$$\mathsf{A}^{-1}\mathsf{A} = \mathsf{A}\mathsf{A}^{-1} = \mathsf{I} = 1 \tag{6.3.1}$$

where I is the unit operator, so that $\mathsf{I}\psi = 1\psi = \psi$ for all states ψ. If B is a second operator we have

$$(\mathsf{B}^{-1}\mathsf{A}^{-1})\,(\mathsf{A}\mathsf{B}) = \mathsf{B}^{-1}(\mathsf{A}^{-1}\mathsf{A})\mathsf{B} = \mathsf{B}^{-1}\mathsf{B} = 1;$$

similarly $(\mathsf{A}^{-1}\mathsf{B})\,(\mathsf{B}^{-1}\mathsf{A}^{-1}) = 1$. Thus

$$(\mathsf{A}\mathsf{B})^{-1} = \mathsf{B}^{-1}\mathsf{A}^{-1} \tag{6.3.2}$$

‡ Further details may be found in Volume 1.

The reader may verify that more generally

$$(\mathsf{AB} \ldots \mathsf{KL})^{-1} = \mathsf{L}^{-1}\mathsf{K}^{-1} \ldots \mathsf{B}^{-1}\mathsf{A}^{-1}. \tag{6.3.3}$$

The *Hermitian conjugate* or *adjoint* $\mathsf{A}^\dagger$ of an operator A satisfies the equation

$$\langle f|\mathsf{A}|g\rangle = \langle \mathsf{A}^\dagger f|g\rangle \tag{6.3.4}$$

for any wave functions f and g. Since

$$\langle f|\mathsf{A}^\dagger|g\rangle = \langle f|\mathsf{A}^\dagger g\rangle = \langle \mathsf{A}^\dagger g|f\rangle^* = \langle g|\mathsf{A}|f\rangle^* = \langle g|\mathsf{A}f\rangle^* = \langle \mathsf{A}f|g\rangle,$$

we see that

$$(\mathsf{A}^\dagger)^\dagger = \mathsf{A}. \tag{6.3.5}$$

Also $\langle f|\mathsf{AB}|g\rangle = \langle \mathsf{A}^\dagger f|\mathsf{B}|g\rangle = \langle \mathsf{B}^\dagger\mathsf{A}^\dagger f|g\rangle$ and so

$$(\mathsf{AB})^\dagger = \mathsf{B}^\dagger\mathsf{A}^\dagger. \tag{6.3.6}$$

The reader may verify that, more generally

$$(\mathsf{AB} \ldots \mathsf{KL})^\dagger = \mathsf{L}^\dagger\mathsf{K}^\dagger \ldots \mathsf{B}^\dagger\mathsf{A}^\dagger. \tag{6.3.7}$$

An operator is *unitary* if it satisfies

$$\mathsf{A}^\dagger\mathsf{A} = \mathsf{A}\mathsf{A}^\dagger = 1, \tag{6.3.8}$$

or equivalently,

$$\mathsf{A}^\dagger = \mathsf{A}^{-1}. \tag{6.3.9}$$

Unitary operators have the important property of preserving scalar product, for

$$\langle \mathsf{A}f|\mathsf{A}g\rangle = \langle \mathsf{A}f|\mathsf{A}|g\rangle = \langle \mathsf{A}^\dagger\mathsf{A}f|g\rangle = \langle f|g\rangle. \tag{6.3.10}$$

In particular, if $\|f\| = \langle f|f\rangle^{\frac{1}{2}}$ as usual,

$$\boxed{\mathsf{A} \text{ is unitary implies } \|\mathsf{A}f\| = \|f\|,} \tag{6.3.11}$$

and so a unitary operator preserves the norm of a vector.

An operator A is *Hermitian* or *self-adjoint* if it satisfies the equation

$$\mathsf{A} = \mathsf{A}^\dagger. \tag{6.3.12}$$

It is well known that the eigenvalues of Hermitian operators are real, and that eigenvectors corresponding to different eigenvalues are orthogonal.

Suppose that H is a Hermitian operator with a complete set of normalized eigenvectors ψ_ν satisfying

$$\boxed{\mathsf{H}\psi_\nu = E_\nu\, \psi_\nu.} \tag{6.3.13}$$

The normalization of the ψ_ν's has the form

$$\boxed{\langle \psi_\nu | \psi_\mu \rangle = \delta_{\mu\nu}} \tag{6.3.14}$$

where $\delta_{\mu\nu}$ is, in general, a product of "Kronecker δ's" and Dirac δ-functions. For example, if we put $\varphi_{\mathbf{k}n} \equiv (2\pi)^{-3/2}\, \Phi_{\mathbf{k}n}$, $\varphi_{\mathbf{l}p} \equiv (2\pi)^{-3/2}\, \Phi_{\mathbf{l}p}$, with the notation of Section 4.1, and use (6.1.9), we see that

$$\langle \varphi_{\mathbf{k}n} | \varphi_{\mathbf{l}p} \rangle = \delta(\mathbf{k}-\mathbf{l})\, \delta_{np}. \tag{6.3.15}$$

The $\varphi_{\mathbf{k}n}$ form a complete set of normalized eigenfunctions of the unperturbed Hamiltonian H_0 of a system of two complex particles, if the χ_n's include the continuum. Any square-integrable wave function Ψ can now be expanded according to the rule

$$\boxed{\Psi = \sum_\nu \psi_\nu\, \langle \psi_\nu | \Psi \rangle} \tag{6.3.16}$$

where Σ_ν represents a summation over discrete values and an integration over the continuum.

If τ is a real number, we may now define the operator $\exp\,(i\mathsf{H}\tau)$ by the usual rule for functions of an operator (see Section 5.2); namely, if Ψ is any square-integrable function, we put

$$\exp\,(i\mathsf{H}\tau)\Psi = \sum_\nu \exp\,(iE_\nu\tau)\psi_\nu\, \langle \psi_\nu | \Psi \rangle. \tag{6.3.17}$$

From (6.3.17) and (5.2.16) we have, if σ is another real number,

$$\begin{aligned} &\exp\,(i\mathsf{H}\tau)\exp\,(i\mathsf{H}\sigma)\Psi \\ &\quad = \sum_\nu \exp\,(iE_\nu\tau)\exp\,(iE_\nu\sigma)\;\psi_\nu\;\langle \psi | \Psi \rangle \\ &\quad = \sum_\nu \exp\,\{iE_\nu(\tau+\sigma)\}\;\psi_\nu\;\langle \psi_\nu | \Psi \rangle \\ &\quad = \exp\,\{i\mathsf{H}(\tau+\sigma)\}\;\Psi; \end{aligned}$$

therefore

$$\exp(i\mathsf{H}\tau)\exp(i\mathsf{H}\sigma) = \exp\{i\mathsf{H}(\tau+\sigma)\}. \tag{6.3.18}$$

That is to say, the same "add the indices" law holds for these operators as for real numbers. One must be careful, however, to note that, if H and H_0 are distinct Hermitian operators, $\exp(i\mathsf{H}\tau)\exp(i\mathsf{H}_0\sigma) \neq \exp(i\mathsf{H}\tau + i\mathsf{H}_0\sigma)$ in general.

If we put $\sigma = -\tau$ or $\tau = -\sigma$ in (6.3.18), we see that

$$\{\exp(i\mathsf{H}\tau)\}^{-1} = \exp(-i\mathsf{H}\tau). \tag{6.3.19}$$

If we put $f(\mathsf{H}) = \exp(i\mathsf{H}\tau)$ in (5.2.12) we find that

$$\{\exp(i\mathsf{H}\tau)\}^{\dagger} = \exp(-i\mathsf{H}\tau). \tag{6.3.20}$$

Equations (6.3.19) and (6.3.20) together show that $\exp(i\mathsf{H}\tau)$ is unitary. Replacement of τ by $-\tau$ in (6.3.19) and (6.3.20) shows that $\exp(-i\mathsf{H}\tau)$ is also unitary, with inverse $\exp(i\mathsf{H}\tau)$.

If Ψ is a wave function depending on τ also, the above results are unaffected. Also if λ is real

$$\begin{aligned}
(\partial/\partial\tau)\,&[\exp(i\lambda\mathsf{H}\tau)\,\Psi(\tau)] \\
&= (\partial/\partial\tau)\sum_{\nu}\exp(i\lambda E_\nu\tau)\,\psi_\nu\,\langle\psi_\nu|\Psi(\tau)\rangle \\
&= \sum_{\nu} i\lambda E_\nu\exp(i\lambda E_\nu\tau)\,\psi_\nu\,\langle\psi_\nu|\Psi(\tau)\rangle \\
&\quad + \sum_{\nu}\exp(i\lambda E_\nu\tau)\,\psi_\nu\left\langle\psi_\nu\middle|\frac{\partial\Psi(\tau)}{\partial\tau}\right\rangle \\
&= i\lambda\mathsf{H}\exp(i\lambda\mathsf{H}\tau)\,\Psi(\tau) + \exp(i\lambda\mathsf{H}\tau)\,\frac{\partial\Psi(\tau)}{\partial\tau}
\end{aligned}$$

so that

$$\frac{\partial}{\partial\tau}\{\exp(i\lambda\mathsf{H}\tau)\,\Psi(\tau)\} = i\lambda\mathsf{H}\exp(i\lambda\mathsf{H}\tau)\,\Psi(\tau) + \exp(i\lambda\mathsf{H}\tau)\,\frac{\partial\Psi(\tau)}{\partial\tau}. \tag{6.3.21}$$

The result (6.3.21) is identical with the result of elementary calculus if H is treated as a constant number.

If we have a sum of operators the reader may verify that the Hermitian conjugate is given by

$$(\mathsf{A}+\mathsf{B}+\ldots+\mathsf{K})^{\dagger} = \mathsf{A}^{\dagger}+\mathsf{B}^{\dagger}+\ldots+\mathsf{K}^{\dagger}. \tag{6.3.22}$$

Suppose now that we have an operator $\mathsf{A}(\tau)$ which depends on a real variable τ. Since an integral is the limit of a sum, it follows that

$$\{\int_{\tau_1}^{\tau_2} A(\tau)\, d\tau\}^{\dagger} = \int_{\tau_1}^{\tau_2} A^{\dagger}(\tau)\, d\tau. \tag{6.3.23}$$

The reader should be able to prove for himself that any function $f(\mathsf{H})$ of H is linear; in other words, if ψ_1 and ψ_2 are two state vectors and c_1, c_2 are two complex numbers, then

$$f(\mathsf{H})\,(c_1\psi_1+c_2\psi_2) = c_1 f(\mathsf{H})\psi_1+c_2 f(\mathsf{H})\psi_2. \tag{6.3.24}$$

In particular, the operator $\exp(-i\mathsf{H}\tau/\hbar)$ is linear.

6.4. The Schrödinger and Heisenberg pictures

In order to take up the evolution of the scattering process we must first consider the evolution in time of general quantum mechanical systems. In quantum mechanics there are two ways of dealing with time-dependence. One way is the so-called *Schrödinger picture*, in which the states evolve in time, whilst the operators are (generally) fixed; the other way is the *Heisenberg picture*, in which the states are fixed and the operators change in time.

Let us consider the first of these. In this picture a state evolves according to the equation

$$i\hbar\frac{\partial\psi}{\partial t} = \mathsf{H}\psi. \tag{6.4.1}$$

A solution of this is

$$\boxed{\psi(t) = \exp\{-i\mathsf{H}(t-t_0)/\hbar\}\,\psi(t_0).} \tag{6.4.2}$$

In the first place ψ takes the value $\psi(t_0)$ at $t = t_0$, as it should; secondly, we have by (6.3.18) and (6.3.21)

$$\begin{aligned} i\hbar\frac{\partial\psi}{\partial t} &= i\hbar\frac{\partial}{\partial t}\exp(-i\mathsf{H}t/\hbar)\,[\exp(i\mathsf{H}t_0/\hbar)\,\psi(t_0)] \\ &= (i\hbar)\left(-\frac{i}{\hbar}\right)\mathsf{H}\exp(-i\mathsf{H}t/\hbar)\exp(i\mathsf{H}t_0/\hbar)\,\psi(t_0) \end{aligned}$$

$$+i\hbar \exp(-i\mathsf{H}t/\hbar)\frac{\partial}{\partial t}[\exp(i\mathsf{H}t_0/\hbar)\,\psi(t_0)]$$

$$= \mathsf{H}\exp\{-i\mathsf{H}(t-t_0)/\hbar\}\,\psi(t_0)$$

$$= \mathsf{H}\psi, \tag{6.4.3}$$

and so (6.4.1) is satisfied.

Equation (6.4.2) is in many ways more fundamental than (6.4.1). For if the wave function is given at time t_0, and the system has a given Hamiltonian H, we can regard (6.4.2) as our fundamental postulate for calculating ψ at any other time t. Equation (6.4.1) is then automatically satisfied if the function is differentiable with respect to time, and we shall assume that this is the case.

In the last section we saw that operators such as $\exp\{-i\mathsf{H}(t-t_0)/\hbar\}$ are unitary, and therefore conserve norm. This is essential for quantum theory to make sense, for it means that $\|\psi(t)\| = \|\psi(t_0)\|$. Wave packets are always normalized to unity since the system must have some configuration, and this must hold for all times. Hence we require that $\|\psi(t)\| = 1$ always; in other words $\|\psi(t)\| = \|\psi(t_0)\|$ for all t and t_0.

To obtain the Heisenberg picture, we transform each wave function $\psi(t)$ into another function $\psi_H(t)$ by means of the unitary transformation

$$\psi_H(t) = \exp(i\mathsf{H}t/\hbar)\,\psi(t). \tag{6.4.4}$$

By application of (6.3.21) to (6.4.4) and then use of (6.4.1)

$$\frac{\partial\psi_H}{\partial t} = \frac{i}{\hbar}\mathsf{H}\exp(i\mathsf{H}t/\hbar)\psi + \exp(i\mathsf{H}t/\hbar)\frac{\partial\psi}{\partial t}$$

$$= \frac{i}{\hbar}\mathsf{H}\exp(i\mathsf{H}t/\hbar)\psi + \exp(i\mathsf{H}t/\hbar)\frac{1}{i\hbar}\mathsf{H}\psi$$

$$= \frac{i}{\hbar}[\mathsf{H}\exp(i\mathsf{H}t/\hbar) - \exp(i\mathsf{H}t/\hbar)\mathsf{H}]\psi$$

$$= 0, \tag{6.4.5}$$

the last step following from (5.2.18). Thus ψ_H is constant.

Let A be any operator, and A_H be the corresponding operator in the Heisenberg picture. Then if ψ is any wave function and ψ_H the corresponding function in the Heisenberg picture, A_H maps $\psi_H(t)$ into the wave function corresponding to $\exp(i\mathsf{H}t/\hbar)\,\mathsf{A}\psi(t)$, and so

$$\mathsf{A}_H\psi_H(t) = \exp(i\mathsf{H}t/\hbar)\,\mathsf{A}\psi(t). \tag{6.4.6}$$

If we operate on (6.4.4) to the left with $\exp(-i\mathsf{H}t/\hbar)$ and use (6.3.19) we see that

$$\exp(-i\mathsf{H}t/\hbar)\,\psi_H(t) = \psi(t), \tag{6.4.7}$$

and so if we substitute for $\psi(t)$ from (6.4.7) into (6.4.6) we have

$$\mathsf{A}_H\psi_H(t) = \exp(i\mathsf{H}t/\hbar)\mathsf{A}\exp(-i\mathsf{H}t/\hbar)\,\psi_H(t). \tag{6.4.8}$$

Since ψ_H is arbitrary, it follows that

$$\mathsf{A}_H = \exp(i\mathsf{H}t/\hbar)\mathsf{A}\exp(-i\mathsf{H}t/\hbar). \tag{6.4.9}$$

The reader may deduce for himself from (6.4.9) that, if A is independent of time,

$$i\hbar\frac{\partial \mathsf{A}_H}{\partial t} = \mathsf{A}_H\mathsf{H} - \mathsf{H}\mathsf{A}_H = [\mathsf{A}_H, \mathsf{H}]; \tag{6.4.10}$$

the well-known Heisenberg equation of motion for the operators in the Heisenberg picture (Volume 1, Section 5.6).

6.5. The interaction picture

For the purposes of scattering theory neither the Schrödinger nor the Heisenberg pictures are particularly useful. Instead we use the *interaction picture*, where we transform each wave function $\psi(t)$ in the Schrödinger picture into a corresponding function $\psi_I(t)$ according to

$$\boxed{\psi_I(t) = \exp(i\mathsf{H}_0t/\hbar)\,\psi(t).} \tag{6.5.1}$$

ψ_I is the function in the "interaction picture" corresponding to $\psi(t)$ in the Schrödinger picture.

By following the arguments of Section 6.4 the reader may show that the operator A in the Schrödinger picture corresponds to the operator A_I in the interaction picture according to

$$\mathsf{A}_I = \exp(i\mathsf{H}_0t/\hbar)\mathsf{A}\exp(-i\mathsf{H}_0t/\hbar). \tag{6.5.2}$$

In particular, the interaction potential V transforms into a time-dependent potential $V_I(t)$ given by

$$\boxed{V_I(t) = \exp(i\mathsf{H}_0t/\hbar)V\exp(-i\mathsf{H}_0t/\hbar).} \tag{6.5.3}$$

Let us now define an *evolution operator* $\mathsf{U}(t, t_0)$, in the interaction picture,‡ by the equation

$$\boxed{\psi_I(t) = \mathsf{U}(t, t_0)\, \psi_I(t_0).} \tag{6.5.4}$$

If we operate on (6.5.1) to the left with $\exp(-i\mathsf{H}_0 t/\hbar)$ and use (6.3.19) with $\mathsf{H} = \mathsf{H}_0$, $\tau = t_0/\hbar$, we see that

$$\psi(t) = \exp(-i\mathsf{H}_0 t/\hbar)\, \psi_I(t), \tag{6.5.5}$$

and the same result holds with $t = t_0$. Hence (6.4.2) becomes

$$\exp(-i\mathsf{H}_0\, t/\hbar)\, \psi_I(t) = \exp[-i\mathsf{H}(t-t_0)/\hbar] \exp(-i\mathsf{H}_0 t_0/\hbar)\, \psi_I(t_0), \tag{6.5.6}$$

and if we operate on (6.5.6) to the left with $\exp(i\mathsf{H}_0 t/\hbar)$ we have

$$\psi_I(t) = \exp(i\mathsf{H}_0 t/\hbar) \exp[-i\mathsf{H}(t-t_0)/\hbar] \exp(-i\mathsf{H}_0 t_0/\hbar)\, \psi_I(t_0). \tag{6.5.7}$$

Comparison of (6.5.4) and (6.5.7) enables us to conclude that

$$\boxed{\mathsf{U}(t, t_0) = \exp(i\mathsf{H}_0 t/\hbar) \exp[-i\mathsf{H}(t-t_0)/\hbar] \exp(-i\mathsf{H}_0 t_0/\hbar).} \tag{6.5.8}$$

Certain properties of $\mathsf{U}(t, t_0)$ follow immediately from (6.5.8). In the first place if we put $t = t_0$ and use (6.3.19) we see that

$$\mathsf{U}(t_0, t_0) = 1, \tag{6.5.9}$$

while application of (6.3.19) and (6.3.18) gives

$$\mathsf{U}(t, t_1)\, \mathsf{U}(t_1, t_0) = \mathsf{U}(t, t_0). \tag{6.5.10}$$

(6.5.9) and (6.5.10) imply

$$\mathsf{U}(t, t_0)\, \mathsf{U}(t_0, t) = \mathsf{U}(t, t) = 1 = \mathsf{U}(t_0, t)\, \mathsf{U}(t, t_0), \tag{6.5.11}$$

so that $\mathsf{U}(t, t_0)$ is the inverse of $\mathsf{U}(t_0, t)$. If we apply (6.3.7) and (6.3.20) to (6.5.8) we see that

$$\mathsf{U}^\dagger(t, t_0) = \mathsf{U}(t_0, t), \tag{6.5.12}$$

and since $\mathsf{U}(t_0, t)$ is the inverse of $\mathsf{U}(t, t_0)$ it follows that the evolution

‡ In Volume 1, $U(t, t_0)$ was defined with reference to the Schrödinger picture: the interaction picture is more useful, however, in discussing the effects of perturbations.

operator is unitary. Since $\exp(i\mathsf{H}_0 t/\hbar)$ is also unitary, we see from (6.5.1) that $\|\psi_I(t)\| = \|\psi(t)\|$, and norms are preserved in the interaction picture as well.

The great advantage of the interaction picture is that the time-dependent equation (6.4.1) now takes a simpler form. For if we differentiate (6.5.1) partially with respect to t remembering that H_0 can be formally treated as a constant number [see (6.3.21)] we obtain

$$\frac{\partial\psi_I}{\partial t} = \left(\frac{i\mathsf{H}_0}{\hbar}\right)\exp(i\mathsf{H}_0 t/\hbar)\psi + \exp(i\mathsf{H}_0 t/\hbar)\frac{\partial\psi}{\partial t}; \tag{6.5.13}$$

hence by (6.4.1)

$$\begin{aligned} i\hbar\frac{\partial\psi_I}{\partial t} &= -\mathsf{H}_0\exp(i\mathsf{H}_0 t/\hbar)\psi + \exp(i\mathsf{H}_0 t/\hbar)\,\mathsf{H}\psi \\ &= -\mathsf{H}_0\exp(i\mathsf{H}_0 t/\hbar)\psi + \exp(i\mathsf{H}_0 t/\hbar)\,(\mathsf{H}_0 + V)\psi. \end{aligned}$$

Since H_0 and $\exp(i\mathsf{H}_0 t/\hbar)$ are both functions of H_0 they commute, and so

$$i\hbar\frac{\partial\psi_I}{\partial t} = \exp(i\mathsf{H}_0 t/\hbar)\,V\psi. \tag{6.5.14}$$

If we make use of (6.5.1) and (6.5.3) we see that (6.5.14) can be rewritten

$$\boxed{i\hbar\frac{\partial\psi_I}{\partial t} = \Psi_I(t)\,\psi_I(t).} \tag{6.5.15}$$

Equation (6.5.15) is the equation in the interaction picture, and is more convenient to handle than the time-dependent Schrödinger equation (6.4.1).

As in the time-dependent case, it is convenient to convert the differential equation into an integral equation. In this case this is easy, for all we have to do is to integrate both sides of (15) from $t = t_0$ to $t = t$ to obtain

$$\psi_I(t) = \psi_I(t_0) - \frac{i}{\hbar}\int_{t_0}^{t} V_I(\tau)\,\psi_I(\tau)\,d\tau. \tag{6.5.16}$$

Further, if $\psi_I(t)$ satisfies (6.5.16) we can differentiate with respect to t, obtaining (6.5.15). Thus (6.5.15) and (6.5.16) are equivalent.

We can also convert the time-dependent problem into an integral equation satisfied by the operator $\mathsf{U}(t, t_0)$. For by use of (6.5.4) we can rewrite (6.5.16) as

$$\mathsf{U}(t, t_0)\, \psi_I(t_0) = \left\{1 - \frac{i}{\hbar} \int_{t_0}^{t} d\tau\, V_I(\tau)\, \mathsf{U}(\tau, t_0)\right\} \psi_I(t_0), \tag{6.5.17}$$

and since (6.5.17) is valid for any wave function $\psi_I(t_0)$ we deduce the operator integral equation

$$\boxed{\mathsf{U}(t, t_0) = 1 - \frac{i}{\hbar} \int_{t_0}^{t} d\tau\, V_I(\tau)\, \mathsf{U}(\tau, t_0).} \tag{6.5.18}$$

The integral equation (6.5.18) for the evolution operator in the interaction picture may also be written in an alternative form, for if we take its Hermitian conjugate and use (6.3.23), (6.3.6) and (6.5.12) we have

$$\mathsf{U}(t_0, t) = 1 + \frac{i}{\hbar} \int_{t_0}^{t} d\tau\, \mathsf{U}(t_0, \tau)\, V_I^{\dagger}(\tau). \tag{6.5.19}$$

The reader may easily verify from (6.5.3) that $V_I^{\dagger}(t) = V_I(t)$, so that $V_I(t)$ is Hermitian. If we now interchange t and t_0 in (6.5.19) we get

$$\boxed{\mathsf{U}(t, t_0) = 1 + \frac{i}{\hbar} \int_{t}^{t_0} d\tau\, \mathsf{U}(t, \tau)\, V_I(\tau),} \tag{6.5.20}$$

a result which we shall find very useful in solving the time-dependent Schrödinger equation.

6.6. Evolution of the wave packet

After the necessary digressions of the last three sections we are now in a position to return to the main scattering problem. We saw in Sections 6.1 and 6.2 that the wave packet emitted by the source of particles has the form

$$\bar{\varphi}(\mathbf{r}, t) = \int d\mathbf{k}\, C(\mathbf{k})\, \varphi_{\mathbf{k}}(\mathbf{r}) \exp(-iE_k t/\hbar). \tag{6.6.1}$$

For the sake of concreteness the reader may consider $\bar{\varphi}$ to be the wave packet actually produced in a physical laboratory as described in Section 6.2; this moves from the source to the target chamber containing the centre of force.‡ Our arguments will, however, be valid for any

‡ In practice there are many scatterers in the target chamber and we assume that they can be treated as independent.

free wave packet; for as we saw in Section 6.1, there is a time t_0 in the remote past when the probability of the particle being in any finite region is negligible. We shall assume that the potential V vanishes outside a sphere of radius R and centre the origin, and so for times $t < t_0$ the wave packet $\bar{\varphi}$ moves freely, and is therefore given by (6.6.1).

For times subsequent to t_0 the potential V comes into play, and the wave packet develops under the full Hamiltonian $\mathsf{H} = \mathsf{H}_0 + V$. Suppose we denote the wave function of the system at time t by $\bar{\psi}(t)$. Then

$$\bar{\psi}(t) = \bar{\varphi}(t), \quad t \leqq t_0, \tag{6.6.2}$$

and in particular

$$\bar{\psi}(t_0) = \bar{\varphi}(t_0). \tag{6.6.3}$$

Let us transform $\bar{\psi}(t)$ into the function $\bar{\psi}_I(t)$ in the interaction picture by

$$\bar{\psi}_I(t) = \exp\,(i\mathsf{H}_0 t/\hbar)\; \bar{\psi}(t). \tag{6.6.4}$$

If we put $t = t_0$ in (6.6.4) we see that

$$\bar{\psi}_I(t_0) = \exp\,(i\mathsf{H}_0 t_0/\hbar)\; \bar{\varphi}(t_0) \quad \text{(by (6.6.3))},$$

and since $\exp\,(i\mathsf{H}_0 t_0/\hbar)\; \varphi_{\mathbf{k}} = \exp\,(iE_k t_0/\hbar)\; \varphi_{\mathbf{k}}$, this combined with (6.6.1) gives us

$$\bar{\psi}_I(t_0) = \bar{\varphi}(0). \tag{6.6.5}$$

It also follows from (6.6.4) that

$$\bar{\psi}_I(0) = \bar{\psi}(0). \tag{6.6.6}$$

We shall now use these preliminary results to solve the time-dependent problem by first evaluating $\bar{\psi}(0)$, and then $\bar{\psi}(t)$. From (6.5.4) we have

$$\bar{\psi}_I(0) = \mathsf{U}(0, t_0)\; \bar{\psi}_I(t_0), \tag{6.6.7}$$

and so by (6.6.6) and (6.6.5)

$$\bar{\psi}(0) = \mathsf{U}(0, t_0)\; \bar{\varphi}(0). \tag{6.6.8}$$

If we substitute for $\mathsf{U}(0, t_0)$ from (6.5.20) into (6.6.8) we obtain

$$\bar{\psi}(0) - \bar{\varphi}(0) = \frac{i}{\hbar} \int_0^{t_0} d\tau\; \mathsf{U}(0, \tau)\; V_I(\tau)\; \bar{\varphi}(0), \tag{6.6.9}$$

and if we replace $\mathsf{U}(0, \tau)$ and $V_I(\tau)$ by the expressions (6.5.8) and (6.5.3) we obtain

$$\bar{\psi}(0) - \bar{\varphi}(0) = \frac{i}{\hbar} \int_0^{t_0} d\tau\; \exp\,(i\mathsf{H}\tau/\hbar) V \exp\,(-i\mathsf{H}_0\tau/\hbar)\; \bar{\varphi}(0). \tag{6.6.10}$$

We can see from (6.6.1) that (6.6.10) can be rewritten

$$\bar{\psi}(0)-\bar{\varphi}(0) = \frac{i}{\hbar}\int_0^{t_0} d\tau \exp\,(i\mathsf{H}\tau/\hbar)V\,\bar{\varphi}(\tau). \tag{6.6.11}$$

We now introduce a small positive number ε which satisfies $\varepsilon|t_0| \ll \hbar$; this enables us to replace (6.6.11) by

$$\bar{\psi}(0)-\bar{\varphi}(0) = \frac{i}{\hbar}\int_0^{t_0} d\tau \exp\,[i(\mathsf{H}-i\varepsilon)\tau/\hbar]V\,\bar{\varphi}(\tau). \tag{6.6.12}$$

Now by hypothesis when $t \leqq t_0$ there is negligible overlap between the region in which V is non-vanishing and $\bar{\varphi}$, and so $V\bar{\varphi}(\tau) = 0$ if $\tau \leqq t_0$. We can therefore replace the upper integration limit t_0 by $-\infty$; (6.6.12) now becomes, with the aid of (6.6.1),

$$\bar{\psi}(0)-\bar{\varphi}(0) = \frac{i}{\hbar}\int_0^{-\infty} d\tau \int d\mathbf{k}\; C(\mathbf{k}) \exp\,[i(\mathsf{H}-E_k-i\varepsilon)\tau/\hbar]V\,\varphi_{\mathbf{k}}. \tag{6.6.13}$$

The integral is evaluated in Appendix D; if we insert this in (6.6.13) and again use (6.6.1) we have

$$\bar{\psi}(0) = \int d\mathbf{k}\; C(\mathbf{k})\,[\varphi_{\mathbf{k}}+(E_k-\mathsf{H}+i\varepsilon)^{-1}V\,\varphi_{\mathbf{k}}]. \tag{6.6.14}$$

We shall define the Schwinger–Lippmann states $\psi_{\mathbf{k}}^{\pm}$ by

$$\psi_{\mathbf{k}}^{\pm} = \varphi_{\mathbf{k}}+(E_k-\mathsf{H}\pm i\varepsilon)^{-1}V\,\varphi_{\mathbf{k}}. \tag{6.6.15}$$

If we multiply our previous definitions

$$\Psi_{\mathbf{k}}^{\pm} = \Phi_{\mathbf{k}}+(E_k-\mathsf{H}\pm i\varepsilon)^{-1}V\Phi_{\mathbf{k}} \tag{6.6.16}$$

by $(2\pi)^{-3/2}$ we see that

$$\psi_{\mathbf{k}}^{\pm} = (2\pi)^{-3/2}\,\Psi_{\mathbf{k}}^{\pm}, \tag{6.6.17}$$

and so the $\psi_{\mathbf{k}}^{\pm}$ satisfy the integral equations

$$\psi_{\mathbf{k}}^{\pm} = \varphi_{\mathbf{k}}+(E_k-\mathsf{H}_0\pm i\varepsilon)^{-1}V_{\mathbf{k}}\psi_{\mathbf{k}}^{\pm}. \tag{6.6.18}$$

Our result (6.6.14) therefore becomes

$$\bar{\psi}(0) = \int d\mathbf{k}\; C(\mathbf{k})\;\psi_{\mathbf{k}}^{+}, \tag{6.6.19}$$

a result which becomes arbitrarily accurate as $\varepsilon \to 0+$ and $t_0 \to -\infty$ in such a way that $\varepsilon t_0 \to 0$. For example, we could take $\varepsilon = T|t_0|^{-2}$ where T is a positive constant.

Now $\bar{\psi}(0)$ evolves under the full Hamiltonian H into the state $\bar{\psi}(t)$, and so from (6.4.2)

$$\bar{\psi}(t) = \exp(-i\mathsf{H}t/\hbar)\,\bar{\psi}(0). \tag{6.6.20}$$

If we substitute for $\bar{\psi}(0)$ from (6.6.19) into (6.6.20) and remember that $\psi_{\mathbf{k}}^{+}$ is an eigenstate of H with energy E_k we have

$$\boxed{\bar{\psi}(t) = \int d\mathbf{k}\; C(\mathbf{k})\; \psi_{\mathbf{k}}^{+} \exp(-iE_k\, t/\hbar),} \tag{6.6.21}$$

which is the required result. We therefore see that the *scattered packet* is formed from stationary state scattering functions by the same prescription that the *incident packet* is formed from free particle functions —justifying the use of the stationary state solutions, as a convenience, in place of the packets which (although both more "physical" and more rigorous) are more difficult to handle.

6.7. Back evolution of a wave packet

We saw in Section 6.1 that the overlap between any free wave packet and the region of interaction becomes negligible as $t \to +\infty$. Let us consider then the free wave packet

$$\bar{\varphi}'(t) = \int d\mathbf{k}\; D(\mathbf{k})\; \varphi_{\mathbf{k}} \exp(-iE_k\, t/\hbar), \tag{6.7.1}$$

and suppose that t_1 is a time sufficiently far in the future for us to suppose that $\bar{\varphi}'$ has no overlap with the region of interaction for $t \geqq t_1$. We now ask ourselves the question: "What wave function develops into the free wave packet $\bar{\varphi}'$ in the far future?" If $\bar{\psi}(t)$ is the required wave function, we have

$$\bar{\psi}(t) = \bar{\varphi}'(t), \quad t \geqq t_1, \tag{6.7.2}$$

and we now seek $\bar{\psi}(t)$.

The evolution operator $\mathsf{U}(t, t_0)$ transforms wave functions *back* in time as well as forward, and so we have

$$\bar{\psi}_I(0) = \mathsf{U}(0, t_1)\; \bar{\psi}_I(t_1) \tag{6.7.3}$$

where $\bar{\psi}_I(t)$ is the transform of $\bar{\psi}(t)$ in the interaction picture. As in Section 6.6 we may now deduce that

$$\bar{\psi}(0) - \bar{\varphi}'(0) = \frac{i}{\hbar} \int_0^{t_1} d\tau\; \mathsf{U}(0, \tau)\; V_I(\tau)\; \bar{\varphi}'(0), \tag{6.7.4}$$

for the integral equations for $\mathsf{U}(t, t_1)$ upon which this result depends are just as valid when $t < t_1$. As before, this equation can be rewritten

$$\bar{\psi}(0) - \bar{\varphi}'(0) = \frac{i}{\hbar} \int_0^{t_1} d\tau \exp (i\mathsf{H}\tau/\hbar) V \bar{\varphi}'(\tau). \tag{6.7.5}$$

We again introduce a small positive number ε satisfying $\varepsilon|t_1| \ll \hbar$, but now replace (6.7.5) by

$$\bar{\psi}(0) - \bar{\varphi}'(0) = \frac{i}{\hbar} \int_0^{t_1} d\tau \exp [i(\mathsf{H}+i\varepsilon)\, \tau/\hbar] V \bar{\varphi}'(\tau). \tag{6.7.6}$$

Since $V\bar{\varphi}'(\tau) = 0$ for $t \geqq t_1$, we can replace t_1 in (6.7.6) by $+\infty$, and use (6.7.1) and the results of Appendix D to evaluate the integral, obtaining

$$\bar{\psi}(0) = \int d\mathbf{k} D(\mathbf{k}) \, [\varphi_{\mathbf{k}} + (E_k - \mathsf{H} - i\varepsilon)^{-1} V \varphi_{\mathbf{k}}]. \tag{6.7.7}$$

From (6.6.15) we see that (6.7.7) is equivalent to

$$\bar{\psi}(0) = \int d\mathbf{k}\, D(\mathbf{k})\, \psi_{\mathbf{k}}^{-}. \tag{6.7.8}$$

Since $\bar{\psi}(t) = \exp(-i\mathsf{H}t/\hbar)\, \bar{\psi}(0)$ and $\mathsf{H}\psi_{\mathbf{k}}^{-} = E_k\, \psi_{\mathbf{k}}^{-}$, (6.7.8) gives us the result

$$\boxed{\bar{\psi}(t) = \int d\mathbf{k}\, D(\mathbf{k})\, \psi_{\mathbf{k}}^{-} \exp(-iE_k\, t/\hbar).} \tag{6.7.9}$$

Thus the wave packet $\bar{\varphi}'(t)$ develops *from* the wave function with the same momentum amplitude, but with the eigenstates normalized to incoming wave-boundary conditions.

CHAPTER 7

THE SCATTERING MATRIX

7.1. Orthogonality of the scattering states

In the last chapter we studied the nature of a freely moving wave packet, and its evolution forward (or back) in time under the action of a central force. We found that in the case of forward evolution the component eigenstates $\varphi_{\mathbf{k}}$ of the free Hamiltonian H_0 are replaced, under the action of the potential V, by the Schwinger–Lippmann eigenstates $\psi_{\mathbf{k}}^{+}$ of the total Hamiltonian $\mathsf{H} = \mathsf{H}_0 + V$. In the case of evolution back in time the free wave packet $\bar{\varphi}'(t)$ was seen to evolve from an interacting packet in which the component free eigenstates $\varphi_{\mathbf{k}}$ were replaced by the Schwinger–Lippmann states $\psi_{\mathbf{k}}^{-}$. It will be our purpose in this chapter to use these facts to obtain some fundamental results about the scattering states, to define the "scattering matrix", and to study its properties.

We have seen that the Schwinger–Lippmann states $\psi_{\mathbf{k}}^{\pm}$ are eigenstates of the total Hamiltonian H with eigenvalues E_k. In addition H will probably have bound states, which we shall denote by ψ_n, with energies E_n. We shall now show that the ψ_n and $\psi_{\mathbf{k}}^{+}$, and the ψ_n and $\psi_{\mathbf{k}}^{-}$, are orthogonal.

The orthogonality of the bound states is well known (Volume 1, Chapter 3). Since $E_n < 0$ and $E_k \geqq 0$, it follows that $\langle \psi_{\mathbf{k}}^{\pm} | \psi_n \rangle = 0$, since $\psi_{\mathbf{k}}^{\pm}$ and ψ_n are eigenstates of the same Hermitian operator, but with different eigenvalues. It remains to show that $\langle \psi_{\mathbf{k}}^{\pm} | \psi_{\mathbf{l}}^{\pm} \rangle = 0$ if $\mathbf{l} \neq \mathbf{k}$.

We shall first of all show that $\langle \psi_{\mathbf{k}}^{\pm} | \psi_n \rangle = 0$ by a different method, whose importance lies in its capacity for generalization. Let $\bar{\varphi}(t)$ be the free wave packet discussed in Section 6.1, and ψ_n be a bound state. Then

$$\lim_{t \to -\infty} \langle \bar{\varphi}(t) | \psi_n \exp(-iE_n t/\hbar \rangle = 0; \tag{7.1.1}$$

for ψ_n is negligible outside some finite region and as $t \to -\infty$ there is negligible overlap between $\bar{\varphi}(t)$ and any finite region.

Under the Hamiltonian H $\bar{\varphi}(t)$ develops into $\bar{\psi}(t)$ as shown in Section 6.6, while $\psi_n \exp(-E_n t/\hbar)$ remains unperturbed, being a solution of the full time-dependent equation (6.4.1). Also the evolution operator

$\exp[-i\mathsf{H}(t-t_0)/\hbar]$ is unitary, and so preserves scalar products [see (6.3.10)]. Hence

$$\begin{aligned}\langle\bar{\psi}(0)|\psi_n\rangle &= \langle\exp[-i\mathsf{H}(0-t)/\hbar]\,\bar{\psi}(t)|\,\exp[-i\mathsf{H}(0-t)/\hbar]\,\psi_n\exp(-iE_n\,t/\hbar)\rangle\\ &= \langle\exp(i\mathsf{H}\,t/\hbar)\,\bar{\psi}(t)|\,\exp(i\mathsf{H}t/\hbar)\,\psi_n\exp(-iE_nt/\hbar)\rangle\\ &= \langle\bar{\psi}(t)|\psi_n\exp(-iE_nt/\hbar)\rangle\\ &= \langle\bar{\varphi}(t)|\psi_n\exp(-iE_nt/\hbar)\rangle,\quad (t\leqq t_0),\end{aligned}\tag{7.1.2}$$

where the last step follows from (6.6.2). If we let $t\to-\infty$ in (7.1.2) and use (7.1.1) we see that

$$\langle\bar{\psi}(0)|\psi_n\rangle = 0,\tag{7.1.3}$$

and so by (6.6.19)

$$\int d\mathbf{k}\;C(\mathbf{k})\;\langle\psi_{\mathbf{k}}^{+}|\psi_n\rangle = 0.\tag{7.1.4}$$

$C(\mathbf{k})$ is arbitrary, however, and so

$$\langle\psi_{\mathbf{k}}^{+}|\psi_n\rangle = 0.\tag{7.1.5}$$

By considering the results of Section 6.7 the reader should also be able to prove in a similar way that

$$\langle\psi_{\mathbf{k}}^{-}|\psi_n\rangle = 0.\tag{7.1.6}$$

Let us now consider the wave packets

$$\bar{\varphi}_1(t) = \int d\mathbf{k}\;C_1(\mathbf{k})\;\varphi_{\mathbf{k}}\exp(-iE_kt/\hbar),\tag{7.1.7}$$

$$\bar{\varphi}_2(t) = \int d\mathbf{k}\;C_2(\mathbf{k})\;\varphi_{\mathbf{k}}\exp(-iE_kt/\hbar).\tag{7.1.8}$$

These move freely in the remote past, and then develop into the wave functions

$$\bar{\psi}_1(t) = \int d\mathbf{k}\;C_1(\mathbf{k})\;\psi_{\mathbf{k}}^{+}\exp(-iE_kt/\hbar),\tag{7.1.9}$$

$$\bar{\psi}_2(t) = \int d\mathbf{k}\;C_2(\mathbf{k})\;\psi_{\mathbf{k}}^{+}\exp(-iE_kt/\hbar).\tag{7.1.10}$$

We therefore have

$$\begin{aligned}\langle\bar{\psi}_1(0)|\bar{\psi}_2(0)\rangle &= \langle\exp(i\mathsf{H}t/\hbar)\,\bar{\psi}_1(t)|\,\exp(i\mathsf{H}t/\hbar)\,\bar{\psi}_2(t)\rangle\\ &= \langle\bar{\psi}_1(t)|\bar{\psi}_2(t)\rangle \underset{t\to-\infty}{\sim} \langle\bar{\varphi}_1(t)|\bar{\varphi}_2(t)\rangle;\end{aligned}\tag{7.1.11}$$

but by (7.1.7) and (7.1.8)

$$\begin{aligned}\langle\bar{\varphi}_1(t)|\bar{\varphi}_2(t)\rangle &= \int d\mathbf{k}\int d\mathbf{l}\;C_1^*(\mathbf{k})\;C_2(\mathbf{l})\exp[i(E_k-E_l)\,t/\hbar]\;\langle\varphi_{\mathbf{k}}|\varphi_{\mathbf{l}}\rangle\\ &= \int d\mathbf{k}\int d\mathbf{l}\;C_1^*(\mathbf{k})\;C_2(\mathbf{l})\;\delta(\mathbf{k}-\mathbf{l}).\end{aligned}\tag{7.1.12}$$

If we substitute for $\bar{\psi}_1(0)$ and $\bar{\psi}_2(0)$ from (7.1.9) and (7.1.10) into

(7.1.11) and use (7.1.12) we see that

$$\int d\mathbf{k}\; C_1^*(\mathbf{k}) \int d\mathbf{l}\; C_2(\mathbf{l})\; \langle \psi_{\mathbf{k}}^{+} | \psi_{\mathbf{l}}^{\pm} \rangle = \int d\mathbf{k}\; C_1^*(\mathbf{k}) \int d\mathbf{l}\; C_2(\mathbf{l})\; \delta(\mathbf{k}-\mathbf{l}). \tag{7.1.13}$$

Now $C_1^*(\mathbf{k})$ is arbitrary, and so (7.1.13) implies

$$\int d\mathbf{l}\; C_2(\mathbf{l})\; \langle \psi_{\mathbf{k}}^{+} | \psi_{\mathbf{l}}^{+} \rangle = \int d\mathbf{l}\; C_2(\mathbf{l})\; \delta(\mathbf{k}-\mathbf{l}); \tag{7.1.14}$$

but $C_2(\mathbf{l})$ is also arbitrary, and so

$$\langle \psi_{\mathbf{k}}^{+} | \psi_{\mathbf{l}}^{+} \rangle = \delta(\mathbf{k}-\mathbf{l}). \tag{7.1.15}$$

By a precisely similar argument which considers the evolution back in time of the wave packets (7.1.7) and (7.1.8) we may also show that

$$\langle \psi_{\mathbf{k}}^{-} | \psi_{\mathbf{l}}^{-} \rangle = \delta(\mathbf{k}-\mathbf{l}). \tag{7.1.16}$$

Since the bound states obey the orthonormality condition

$$\langle \psi_m | \psi_n \rangle = \delta_{mn}, \tag{7.1.17}$$

(7.1.5) and (7.1.15) show that the bound states and continuum states $\psi_{\mathbf{k}}^{+}$ together form an orthonormal set, while (7.1.6) and (7.1.16) show that the ψ_n and ψ_k^{-} together form a second orthonormal set. The question of their completeness will be considered in the next section.

7.2. The completeness theorem

Let us recall the physical situation which we are attempting to describe. The target particles in the target chamber are regarded as infinitely massive, and we regard each scattering event as independent. We are therefore looking at the case of a single particle moving in the field of a central potential $V(\mathbf{r})$. The particle is therefore either free or bound. If it is bound, it remains so for all time, for there is no external agency to alter its condition; but if it is free, it will be at an infinite distance from the centre of force in the infinite past or the infinite future. We can crystallize these ideas in the following postulate:

> POSTULATE. If a physical system consists of a particle moving in the field of a central potential of finite range, and the system is observed in the remote past or the remote future, the only possible observations are: (i) the particle moving freely; (ii) the particle bound to the centre of force. (7.2.1)

It follows from the basic laws of quantum mechanics that if $\bar{\psi}(t)$ is the wave function at time t

$$\bar{\psi}(t) \underset{t\to-\infty}{\sim} \sum_n C_n\psi_n \exp(-iE_nt/\hbar)+\int d\mathbf{k}\; C(\mathbf{k})\; \varphi_{\mathbf{k}} \exp(-iE_kt/\hbar), \tag{7.2.2}$$

$$\bar{\psi}(t) \underset{t\to+\infty}{\sim} \sum_n D_n\psi_n \exp(-iE_nt/\hbar)+\int d\mathbf{k}\; D(\mathbf{k})\; \varphi_{\mathbf{k}} \exp(-iE_kt/\hbar). \tag{7.2.3}$$

Suppose that t_0 is a sufficiently negative time for the limit (7.2.2) to be effectively attained when $t = t_0$. Then

$$\bar{\psi}(t_0) = \sum_n C_n\psi_n \exp(-iE_nt_0/\hbar)+\int d\mathbf{k}\; C(\mathbf{k})\; \varphi_{\mathbf{k}} \exp(-iE_kt_0/\hbar), \tag{7.2.4}$$

and by (6.6.1) this may be written

$$\bar{\psi}(t_0) = \sum_n C_n\psi_n \exp(-iE_nt_0/\hbar)+\bar{\varphi}(t_0). \tag{7.2.5}$$

If we operate on (7.2.5) to the left with the evolution operator $\exp[-i\mathsf{H}(t-t_0)/\hbar]$, and remember that this is linear, we have

$$\bar{\psi}(t) = \sum_n C_n\psi_n \exp(-iE_nt/\hbar)+\exp[-i\mathsf{H}(t-t_0)/\hbar]\; \bar{\varphi}(t_0). \tag{7.2.6}$$

Since $\bar{\varphi}(t_0)$ evolves in the way described in Section 6.6, we deduce that

$$\bar{\psi}(t) = \sum_n C_n\psi_n \exp(-iE_nt/\hbar)+\int d\mathbf{k}\; C(\mathbf{k})\; \psi_{\mathbf{k}}^{+} \exp(-iE_kt/\hbar). \tag{7.2.7}$$

In particular, when $t = 0$

$$\boxed{\bar{\psi}(0) = \sum_n C_n\psi_n+\int d\mathbf{k}\; C(\mathbf{k})\; \psi_{\mathbf{k}}^{+}.} \tag{7.2.8}$$

Now $\bar{\psi}(0)$ can be any arbitrary wave function, and (7.2.8) shows that this can be expanded in terms of the bound states ψ_n and the scattering states $\psi_{\mathbf{k}}^{+}$ with outgoing wave conditions. It therefore follows from the properties proved in Section 7.1 that the states ψ_n combined with the states $\psi_{\mathbf{k}}^{+}$ form a complete set of orthonormal eigenstates of the total Hamiltonian H. In exactly the same way, by consideration of the condition (7.2.3) and the results of Sections 6.7 and 7.1 we can see that the ψ_n combined with the $\psi_{\mathbf{k}}^{-}$ form a second complete orthonormal set of eigenstates of H.

Since $C_n = \langle\psi_n|\bar{\psi}(0)\rangle$ whether the $\psi_{\mathbf{k}}^{+}$ or the $\psi_{\mathbf{k}}^{-}$ are used, the bound states span one part of Hilbert space, and the $\psi_{\mathbf{k}}^{+}$ or $\psi_{\mathbf{k}}^{-}$ span the other part. These two parts are orthogonal, in the sense that any element of one part is clearly orthogonal to any element of the other. The part spanned by the $\psi_{\mathbf{k}}^{+}$ (or $\psi_{\mathbf{k}}^{-}$) is known as the "continuum" part of Hilbert space.

7.3. The scattering matrix

The behaviour of the system in the remote past or remote future is given by (7.2.2) or (7.2.3). By following the *forward* evolution in time of $\bar{\psi}(t)$ from the state (7.2.2) we have seen that [see (7.2.7)]

$$\bar{\psi}(t) = \sum_{n} C_n\psi_n \exp(-iE_nt/\hbar) + \int d\mathbf{k}\; C(\mathbf{k})\; \psi_{\mathbf{k}}^{+} \exp(-iE_kt/\hbar), \tag{7.3.1}$$

and in the same way by following the backward evolution in time of (7.2.3) we see that

$$\bar{\psi}(t) = \sum_{n} D_n\psi_n \exp(-iE_nt/\hbar) + \int d\mathbf{k}\; D(\mathbf{k})\; \psi_{\mathbf{k}}^{-} \exp(-iE_{\mathbf{k}}t/\hbar). \tag{7.3.2}$$

The eigenstates ψ_n and $\psi_{\mathbf{k}}^{+}$ make up a complete orthonormal set, so let us denote the general member of this set by ψ_{ν}^{+}. Then (7.3.1) can be abbreviated to

$$\bar{\psi}(t) = \sum_{\nu} C_{\nu}\; \psi_{\nu}^{+} \exp(-iE_{\nu}t/\hbar), \tag{7.3.3}$$

while the orthogonality properties (7.1.5), (7.1.15), and (7.1.17) may be abbreviated by the single formula

$$\langle\psi_{\mu}^{+}|\psi_{\nu}^{+}\rangle = \delta_{\mu\nu}. \tag{7.3.4}$$

At the same time we can denote the general member of the complete orthonormal set ψ_n, $\psi_{\mathbf{k}}^{-}$ by ψ_{ν}^{-}, so that (7.3.2) becomes

$$\bar{\psi}(t) = \sum_{\nu} D_{\nu}\; \psi_{\nu}^{-} \exp(-iE_{\nu}t/\hbar), \tag{7.3.5}$$

while (7.1.6), (7.1.16) and (7.1.17) may be summarized by

$$\langle\psi_{\mu}^{-}|\psi_{\nu}^{-}\rangle = \delta_{\mu\nu}. \tag{7.3.6}$$

It follows immediately from (7.3.3) and (7.3.5) that

$$\sum_{\nu} D_{\nu}\; \psi_{\nu}^{-} = \sum_{\nu} C_{\nu}\; \psi_{\nu}^{+}, \tag{7.3.7}$$

since both sides equal $\bar{\psi}(0)$. If we take the scalar product of (7.3.7) on the left with ψ_{μ}^{-} and use (7.3.6) we see that

$$D_{\mu} = \sum_{\nu} \langle\psi_{\mu}^{-}|\psi_{\nu}^{+}\rangle\; C_{\nu}. \tag{7.3.8}$$

Let us put

$$\boxed{S_{\mu,\nu} \equiv \langle \psi_\mu^- | \psi_\nu^+ \rangle;} \tag{7.3.9}$$

then we can rewrite (7.3.8) as

$$\boxed{D_\mu = \sum_\nu S_{\mu,\nu} C_\nu.} \tag{7.3.10}$$

Equation (7.3.10) has the form of a matrix equation. The set of elements C_ν forms a vector whose components are the probability amplitudes for the possible states of the system in the remote past. Similarly the elements D_μ form a vector whose components are the probability amplitudes of the possible states in the far future. The set of numbers $S_{\mu,\nu}$ make up a matrix which transforms the vector C_ν describing the initial state of the system into the vector D_μ describing the final state of the system; we can call this the "scattering matrix".

The elements of the scattering matrix between bound states are

$$S_{p,n} = \langle \psi_p | \psi_n \rangle = \delta_{pn}; \tag{7.3.11}$$

between bound and continuum states they are

$$S_{p,\mathbf{k}} = \langle \psi_p | \psi_\mathbf{k}^+ \rangle = 0, \tag{7.3.12}$$

$$S_{\mathbf{l},n} = \langle \psi_\mathbf{l}^- | \psi_n \rangle = 0, \tag{7.3.13}$$

and between continuum states they are

$$S_{\mathbf{l},\mathbf{k}} = \langle \psi_\mathbf{l}^- | \psi_\mathbf{k}^+ \rangle. \tag{7.3.14}$$

Suppose we consider the scattering problem alone, so that $C_n = 0$ for all bound states ψ_n. (7.3.10) now gives, from (7.3.12),

$$D_p = \int d\mathbf{k}\, S_{p,\mathbf{k}}\, C(\mathbf{k}) = 0 \tag{7.3.15}$$

and so there are no bound states finally. This says no more than that a free particle cannot be captured into a bound state. The only way that this can happen is for the particle to emit energy in the form of a photon, in which case the final state would contain a photon as well as a particle contrary to the postulate (7.2.1). The effect of this postulate is to exclude interaction with the electromagnetic field; such interactions, which are fundamental to photochemistry, for example, will not be considered in this book.

We also obtain from (7.3.10) and (7.3.13) that

$$D(\mathbf{l}) = \int d\mathbf{k}\, S_{\mathbf{l},\mathbf{k}}\, C(\mathbf{k}). \tag{7.3.16}$$

The elements $S_{\mathbf{l},\mathbf{k}}$ form a matrix which transforms the vector describing the initial state of the free particle into the vector describing the final state, when the particle is again free. Knowledge of the matrix $S_{\mathbf{l},\mathbf{k}}$ enables us to predict the result of any scattering experiment, and so this matrix, which is a sub-matrix of the matrix (7.3.9), is also known as the scattering matrix.

7.4. Calculation of the scattering matrix

Our object in this section will be to obtain a very famous expression for the elements $S_{\mathbf{l},\mathbf{k}}$ of the scattering matrix, which will be very useful to us in obtaining various results of a practical nature. From (7.3.14) and (7.1.15) we see that

$$\begin{aligned} S_{\mathbf{l},\mathbf{k}} &= \langle \psi_{\mathbf{l}}^{-} | \psi_{\mathbf{k}}^{+} \rangle \\ &= \langle \psi_{\mathbf{l}}^{+} | \psi_{\mathbf{k}}^{+} \rangle + \langle \psi_{\mathbf{l}}^{-} - \psi_{\mathbf{l}}^{+} | \psi_{\mathbf{k}}^{+} \rangle \\ &= \delta(\mathbf{l}-\mathbf{k}) + \langle \psi_{\mathbf{l}}^{-} - \psi_{\mathbf{l}}^{+} | \psi_{\mathbf{k}}^{+} \rangle. \end{aligned} \tag{7.4.1}$$

We must now calculate the second term on the right-hand side of (7.4.1). We shall do this by considering the quantity $\langle \psi_{\mathbf{l}}^{-} - \psi_{\mathbf{l}}^{+} | \bar{\psi}(0) \rangle$ where $\bar{\psi}(t)$ is the wave function of Section 6.6. For from (6.6.15) with $\mathbf{k}=\mathbf{l}$

$$\begin{aligned} \langle \psi_{\mathbf{l}}^{-} - \psi_{\mathbf{l}}^{+} | \bar{\psi}(0) \rangle &= \langle [(E_l - \mathsf{H} - i\varepsilon)^{-1} - (E_l - \mathsf{H} + i\varepsilon)^{-1}]\; V\varphi_{\mathbf{l}} | \bar{\psi}(0) \rangle \\ &= \langle V\varphi_{\mathbf{l}} | (E_l - \mathsf{H} + i\varepsilon)^{-1} - (E - \mathsf{H} - i\varepsilon)^{-1} | \bar{\psi}(0) \rangle, \end{aligned} \tag{7.4.2}$$

and so by (6.6.21)

$$\begin{aligned} &\langle \psi_{\mathbf{l}}^{-} - \psi_{\mathbf{l}}^{+} | \bar{\psi}(0) \rangle \\ &= \int d\mathbf{k}\; C(\mathbf{k})\; \langle V\varphi_{\mathbf{l}} | (E_l - E_k + i\varepsilon)^{-1} - (E_l - E_k - i\varepsilon)^{-1} | \psi_{\mathbf{k}}^{+} \rangle \\ &= \int d\mathbf{k}\; C(\mathbf{k}) \frac{-2i\varepsilon}{(E_l - E_k)^2 + \varepsilon^2} \langle V\varphi_{\mathbf{l}} | \psi_{\mathbf{k}}^{+} \rangle. \end{aligned} \tag{7.4.3}$$

Now we can easily show that

$$\lim_{\varepsilon \to 0+} \frac{\varepsilon}{x^2 + \varepsilon^2} = \pi \delta(x). \tag{7.4.4}$$

For if $x \neq 0$, the left-hand side is zero, and

$$\int_{-\infty}^{+\infty} \frac{\varepsilon\, dx}{x^2 + \varepsilon^2} = \left[\tan^{-1} \frac{x}{\varepsilon} \right]_{x=-\infty}^{x=+\infty} = \pi, \tag{7.4.5}$$

which establishes (7.4.4). Hence (7.4.3) becomes, on use of (6.6.21),

$$\int d\mathbf{k}\, C(\mathbf{k})\, \langle \psi_{\mathbf{l}}^{-} - \psi_{\mathbf{l}}^{+} | \psi_{\mathbf{k}}^{+} \rangle$$
$$= \int d\mathbf{k}\, C(\mathbf{k})\, (-2\pi i)\, \delta(E_l - E_k)\, \langle \varphi_{\mathbf{l}} | V | \psi_{\mathbf{k}}^{+} \rangle. \tag{7.4.6}$$

Since $C(\mathbf{k})$ is arbitrary we conclude that

$$\langle \psi_{\mathbf{l}}^{-} - \psi_{\mathbf{l}}^{+} | \psi_{\mathbf{k}}^{+} \rangle = -2\pi i\, \delta(E_l - E_k)\, \langle \varphi_{\mathbf{l}} | V | \psi_{\mathbf{k}}^{+} \rangle. \tag{7.4.7}$$

If we insert (7.4.7) into (7.4.1) we obtain the result

$$\boxed{S_{\mathbf{l},\mathbf{k}} = \delta(\mathbf{l} - \mathbf{k}) - 2\pi i\, \delta(E_l - E_k)\, \langle \varphi_{\mathbf{l}} | V | \psi_{\mathbf{k}}^{+} \rangle.} \tag{7.4.8}$$

In the special case of scattering by a centre of force (5.5.15) and (5.5.21) show that

$$\langle \Phi_{\mathbf{l}} | \mathsf{T} | \Phi_{\mathbf{k}} \rangle = \langle \Phi_{\mathbf{l}} | V | \Psi_{\mathbf{k}}^{+} \rangle. \tag{7.4.9}$$

If we multiply both sides of (7.4.9) by $(2\pi)^{-3}$ and recall that $\varphi_{\mathbf{k}} = (2\pi)^{-3/2}\Phi_{\mathbf{k}}$. $\psi_{\mathbf{k}}^{+} = (2\pi)^{-3/2}\Psi_{\mathbf{k}}^{+}$, we find that

$$\langle \varphi_{\mathbf{l}} | \mathsf{T} | \varphi_{\mathbf{k}} \rangle = \langle \varphi_{\mathbf{l}} | V | \psi_{\mathbf{k}}^{+} \rangle. \tag{7.4.10}$$

Insertion of (7.4.10) into (7.4.8) gives us the alternative result

$$\boxed{S_{\mathbf{l},\mathbf{k}} = \delta(\mathbf{l} - \mathbf{k}) - 2\pi i\, \delta(E_l - E_k)\, \langle \varphi_{\mathbf{l}} | \mathsf{T} | \varphi_{\mathbf{k}} \rangle.} \tag{7.4.11}$$

7.5. The final wave packet

Let us summarize the results of previous sections. We have seen that the wave function $\bar{\psi}(t)$ which describes the scattering in time of a single structureless particle by a centre of force is given by the expression

$$\bar{\psi}(t) = \int d\mathbf{k}\, C(\mathbf{k})\, \psi_{\mathbf{k}}^{+} \exp(-iE_k t/\hbar). \tag{7.5.1}$$

In the remote past this behaves freely, taking the asymptotic form

$$\bar{\psi}(t) \underset{t \to -\infty}{\sim} \bar{\varphi}(t) \equiv \int d\mathbf{k}\, C(\mathbf{k})\, \varphi_{\mathbf{k}} \exp(-iE_k t/\hbar). \tag{7.5.2}$$

It also behaves freely in the remote future, and therefore has the asymptotic behaviour

$$\bar{\psi}(t) \underset{t \to +\infty}{\sim} \bar{\varphi}'(t) \equiv \int d\mathbf{l}\, D(\mathbf{l})\, \varphi_{\mathbf{l}} \exp(-iE_l t/\hbar). \tag{7.5.3}$$

$\bar{\varphi}(t)$ is the initial wave packet, $\bar{\varphi}'(t)$ the final wave packet, while $C(\mathbf{k})$ and $D(\mathbf{l})$ are the initial and final momentum amplitudes. The momentum amplitudes are related by the expression (7.3.16); that is by

$$D(\mathbf{l}) = \int d\mathbf{k}\, S_{\mathbf{l},\mathbf{k}}\, C(\mathbf{k}), \tag{7.5.4}$$

where $S_{\mathbf{l},\mathbf{k}}$ is the scattering matrix. In this section we shall first investigate the nature of the final momentum amplitude $D(\mathbf{l})$ and then we shall investigate the nature of the final wave function $\bar{\varphi}'(t)$ in the case of an actual scattering experiment. $D(\mathbf{l})$ is the final state in the "momentum representation", and $\bar{\varphi}'(t)$ is the final state in the "position" or Schrödinger representation (Volume 1, Chapter 5).

If we substitute for $S_{\mathbf{l},\mathbf{k}}$ from (7.4.11) into (7.5.4) we see that

$$D(\mathbf{l}) = C(\mathbf{l}) + \Delta(\mathbf{l}), \tag{7.5.5}$$

where

$$\Delta(\mathbf{l}) \equiv -2\pi i \int d\mathbf{k}\, \delta(E_l - E_k)\, \langle \varphi_{\mathbf{l}} | \mathsf{T} | \varphi_{\mathbf{k}} \rangle\, C(\mathbf{k}). \tag{7.5.6}$$

$\Delta(\mathbf{l})$ is the change produced in the momentum amplitude by the collision. If we express $\mathbf{k}$ in terms of spherical polar coordinates in $\mathbf{k}$-space, so that $d\mathbf{k} = k^2\, dk d\hat{\mathbf{k}}$ where $d\hat{\mathbf{k}}$ is an element of solid angle in the direction of the unit vector $\hat{\mathbf{k}}$, and carry out the integration over k in (7.5.6) we obtain

$$\Delta(\mathbf{l}) = -2\pi i\, \hbar^{-2}\, \mu k \int d\hat{\mathbf{k}}\, \langle \varphi_{\mathbf{l}} | \mathsf{T} | \varphi_{\mathbf{k}} \rangle\, C(\mathbf{k}), \quad (E_l = E_k). \tag{7.5.7}$$

We may express (7.5.7) in a simpler form if we relate $\langle \varphi_{\mathbf{l}} | \mathsf{T} | \varphi_{\mathbf{k}} \rangle$ to the scattering amplitude $f(\mathbf{k} \rightarrow \mathbf{l})$. In the special case of the scattering of a particle by a centre of force (5.5.21) becomes

$$f(\mathbf{k} \rightarrow \mathbf{l}) = -(\mu/2\pi\hbar^2)\, \langle \Phi_{\mathbf{l}} | \mathsf{T} | \Phi_{\mathbf{k}} \rangle \tag{7.5.8}$$

where the transition operator T is given by (5.5.20). Equation (7.5.8) may be rewritten

$$f(\mathbf{k} \rightarrow \mathbf{l}) = -(4\pi^2\, \mu/\hbar^2)\, \langle \varphi_{\mathbf{l}} | \mathsf{T} | \varphi_{\mathbf{k}} \rangle, \tag{7.5.9}$$

and so (7.5.7) becomes, if we remember that $E_l = E_k$ and so $l = k$,

$$\Delta(\mathbf{l}) = (2\pi)^{-1} i l \int d\hat{\mathbf{k}}\, f(l\hat{\mathbf{k}} \rightarrow \mathbf{l})\, C(l\hat{\mathbf{k}}). \tag{7.5.10}$$

Equation (7.5.10) enables us to draw several conclusions. In the first place the magnitude of $\Delta(\mathbf{l})$ increases with the magnitude of f, so that the amount of scattering increases with the scattering amplitude. In the second place $C(l\hat{\mathbf{k}})$ vanishes unless $|l\hat{\mathbf{k}} - \mathbf{k}_0| = |\mathbf{k} - \mathbf{k}_0| < \Delta k$, where

$$\Delta k = \{(\Delta k_x)^2 + (\Delta k_y)^2 + (\Delta k_z)^2\}^{\frac{1}{2}}, \tag{7.5.11}$$

Δk_x, Δk_y and Δk_z being defined in Section 6.2. Since $|l\hat{\mathbf{k}} - \mathbf{k}_0| \geqq |l - k_0|$, it follows from (7.5.10) that $\Delta(\mathbf{l})$ vanishes if $|l-k_0| \geqq \Delta k$, and so the energy range of the scattered particles is the same as that of the incident beam, which exemplifies the conservation of energy.

Now if observations are to be made, the scattering amplitude must not vary much as $\hat{\mathbf{k}}$ takes values for which $C(l\hat{\mathbf{k}}) \neq 0$. That is to say, the incident beam must be sufficiently well collimated for us to assume that, as the direction $\hat{\mathbf{k}}$ of the incident particle varies over those values for which $C(l\hat{\mathbf{k}}) \neq 0$, or in other words, over the possible directions of the particles of the incident beam, f must remain constant. We can therefore put $\hat{\mathbf{k}} = \hat{\mathbf{k}}_0$, where $\hat{\mathbf{k}}_0$ is a unit vector along Oz, in the scattering amplitude and take it outside the integral in (7.5.10). Thus

$$\Delta(\mathbf{l}) = (2\pi)^{-1} il\, f(l\hat{\mathbf{k}}_0 \rightarrow \mathbf{l}) \int d\hat{\mathbf{k}}\, C(l\hat{\mathbf{k}}). \tag{7.5.12}$$

We shall now investigate the shape of the final wave packet $\bar{\varphi}'(t)$. If we substitute for $D(\mathbf{l})$ from (7.5.5) into (7.5.3) and use (7.5.2) we have

$$\bar{\varphi}'(t) = \bar{\varphi}(t) + \bar{\psi}_s(t) \tag{7.5.13}$$

where

$$\bar{\psi}_s(t) \equiv \int d\mathbf{l}\, \Delta(\mathbf{l})\, \varphi_{\mathbf{l}} \exp(-iE_l t/\hbar). \tag{7.5.14}$$

Thus the final wave packet consists of the sum of the unperturbed initial packet $\bar{\varphi}(t)$ and a scattered wave $\bar{\psi}_s(t)$. We have already investigated the shape of the initial packet, and so it remains to consider $\bar{\psi}_s(t)$.

We can rewrite (7.5.14) as

$$\bar{\psi}_s(\mathbf{r}, t) = (2\pi)^{-3/2} \int d\mathbf{l} |\Delta(\mathbf{l})| \exp[i\mathbf{l}\cdot\mathbf{r} - iE_l t/\hbar + i \arg \Delta(\mathbf{l})]. \tag{7.5.15}$$

Let $\mathbf{l}$ have coordinates (l, ξ, η) relative to a system of spherical polar coordinates with $\mathbf{r}$ as axis; then (7.5.15) becomes

$$\bar{\psi}_s(\mathbf{r}, t) = (2\pi)^{-3/2} \int_0^\infty l^2\, dl \int_0^{2\pi} d\eta \int_0^\pi \sin\xi\, d\xi \times$$
$$\times |\Delta(\mathbf{l})| \exp[ilr\cos\xi - iE_l\, t/\hbar + i \arg \Delta(\mathbf{l})]. \tag{7.5.16}$$

Now since $\hat{\mathbf{r}}$ is the polar axis, $\mathbf{l} = \pm l\hat{\mathbf{r}}$ when $\xi = 0$ or π, and so repeated application of integration by parts to the integral over ξ in (7.5.16), with $\exp(ilr\cos\xi)\sin\xi\, d\xi$ treated as the differential, gives

$$\bar{\psi}_s(\mathbf{r}, t) = (2\pi)^{-3/2} \int_0^\infty l^2\, dl \int_0^{2\pi} d\eta \times$$
$$\times \{(ilr)^{-1} |\Delta(l\hat{\mathbf{r}})| \exp[ilr - iE_l\, t/\hbar + i \arg \Delta(l\hat{\mathbf{r}})]$$
$$-(ilr)^{-1} |\Delta(-l\hat{\mathbf{r}})| \exp[-ilr - iE_l\, t/\hbar + i \arg \Delta(-l\hat{\mathbf{r}})]$$
$$+ 0[(ilr)^{-2}]\}. \tag{7.5.17}$$

If we put $\mathbf{l} = \pm l\hat{\mathbf{r}}$ in (7.5.12) we see that

$$\Delta(\pm l\hat{\mathbf{r}}) = (2\pi)^{-1}\, il\, f(l\hat{\mathbf{k}}_0 \to \pm l\hat{\mathbf{r}}) \int d\hat{\mathbf{k}}\; C(l\hat{\mathbf{k}}), \tag{7.5.18}$$

and this vanishes unless $l \simeq k_0$. Hence in the integrand of (7.5.17) $lr \simeq k_0 r$, and if we consider $\bar{\psi}_s(\mathbf{r}, t)$ only at points distant from the origin so that $k_0 r \gg 1$, and note that the integrand of (7.5.17) is independent of η, we get

$$\bar{\psi}_s(\mathbf{r}, t) \underset{k_0 r \to +\infty}{\sim} \frac{(2\pi)^{-\frac{1}{2}}}{ir} \int_0^\infty l\, dl\, \{|\Delta(l\hat{\mathbf{r}})| \exp\,[ilr - iE_l t/\hbar + i \arg \Delta(l\hat{\mathbf{r}})]$$
$$- |\Delta(-l\hat{\mathbf{r}})| \exp\,[-ilr - iE_l t/\hbar + i \arg \Delta(-l\hat{\mathbf{r}})]\}. \tag{7.5.19}$$

In Section 6.2 we investigated the nature of the initial wave packet $\bar{\varphi}(t)$ by examining the variation of the argument of the exponential over the interval of integration, and we shall use the same approach here. We first note that

$$(\partial/\partial l)\,[-lr - E_l t/\hbar + \arg \Delta(-l\hat{\mathbf{r}})] = -r - \hbar l t/\mu + (\partial/\partial l) \arg \Delta(-l\hat{\mathbf{r}})$$
$$\leqq -\hbar l t/\mu + (\partial/\partial l) \arg \Delta(-l\hat{\mathbf{r}}) \tag{7.5.20}$$

since $r \geqq 0$. This tends to $-\infty$ as $t \to +\infty$, and so the second term on the right-hand side of (7.5.19) tends to zero, due to the oscillation of the exponential. Hence

$$\bar{\psi}_s(\mathbf{r}, t) \underset{k_0 r,\, t \to +\infty}{\sim} \frac{(2\pi)^{-\frac{1}{2}}}{ir} \int_0^\infty l\, dl\, |\Delta(l\hat{\mathbf{r}})| \exp\,[ilr - iE_l t/\hbar + i \arg \Delta(l\hat{\mathbf{r}})], \tag{7.5.21}$$

and so behaves as an outgoing wave. The rate of change of the argument of the exponential in (7.5.21) is

$$(\partial/\partial l)\,[lr - E_l\, t/\hbar + \arg \Delta(l\hat{\mathbf{r}})] = r - \hbar l(t - t'_D)/\mu \tag{7.5.22}$$

where the quantity t'_D is defined by

$$t'_D \equiv \hbar \frac{\partial}{\partial E_l} \arg \Delta(l\hat{\mathbf{r}}) = t'_D(l\hat{\mathbf{r}}). \tag{7.5.23}$$

Now since $\Delta(l\hat{\mathbf{r}})$ vanishes if $|l - k_0| > \Delta k$, the effective range of integration in (7.5.21) is from $l = k_0 - \Delta k$ to $l = k_0 + \Delta k$. Let

$$T_D(\hat{\mathbf{r}}) \equiv \max_{|l-k_0|<\Delta k} t'_D(l\hat{\mathbf{r}}), \tag{7.5.24}$$

$$\tau_D(\hat{\mathbf{r}}) \equiv \min_{|l-k_0|<\Delta k} t'_D(l\hat{\mathbf{r}}). \tag{7.5.25}$$

Then from (7.5.22), if $k_0 r \gg 1$ and $t > T_D(\hat{\mathbf{r}})$, the scattered wave $\bar{\psi}_s(t)$ vanishes if

$$r - \hbar(k_0 - \Delta k)\,[t - T_D(\hat{\mathbf{r}})]/\mu < -n\pi/\Delta k,$$

$$r - \hbar(k_0 + \Delta k)\,[t - \tau_D(\hat{\mathbf{r}})]/\mu > n\pi/\Delta k, \tag{7.5.26}$$

due to the oscillation of the exponential in (7.5.21), n being a large positive integer as usual. As we saw in Section 6.2 the mean velocity v_0 of the original wave packet is given by $v_0 = \hbar k_0/\mu$, while the spread of velocities v in the incident beam is $v_0 - \Delta v < v < v_0 + \Delta v$, where $\Delta v \equiv \hbar \Delta k/\mu$. Hence from (7.5.26), if $\bar{\psi}_s(\mathbf{r}, t) \neq 0$ in a given direction $\hat{\mathbf{r}}$,

$$(v_0 - \Delta v)\,[t - T_D(\hat{\mathbf{r}})] - n\pi/\Delta k < r < (v_0 + \Delta v)\,[t - \tau_D(\hat{\mathbf{r}})] + n\pi/\Delta k. \tag{7.5.27}$$

The shape of the region to which the scattered wave is confined is therefore a thin shell whose thickness is of the order of $2\Delta v t$ for large t, and whose inner and outer surfaces expand with speeds $v_0 - \Delta v$ and $v_0 + \Delta v$ respectively. It therefore expands away from the origin. If $T_D(\hat{\mathbf{r}})$ and $\tau_D(\hat{\mathbf{r}})$ are both independent of $\hat{\mathbf{r}}$, the shell is spherical.

In the special case when $C(\mathbf{k})$ is real (7.5.18) and (7.5.23) show that $t_D' = t_D$ where

$$t_D \equiv \hbar \frac{\partial}{\partial E_l} \arg f(l\hat{\mathbf{k}}_0 \to l\hat{\mathbf{r}}) = t_D(\hat{\mathbf{r}}). \tag{7.5.28}$$

Thus, if t_D is assumed independent of l, $T_D(\hat{\mathbf{r}}) = \tau_D(\hat{\mathbf{r}}) = t_D(\hat{\mathbf{r}})$, and (7.5.27) shows that the central part of the scattered wave is the surface

$$r = v_0[t - t_D(\hat{\mathbf{r}})]. \tag{7.5.29}$$

Thus the particle emerges in the direction $\hat{\mathbf{r}}$ as if it were at the centre O at time $t = t_D\hat{\mathbf{r}}$, where from the work of Section 6.2 we can see that the reality of $C(\mathbf{k})$ implies that $\mathbf{M} = \mathbf{m} = \mathbf{O}$, so that the centre of the free wave packet would have passed O at time $t = 0$. The quantity $t_D(\hat{\mathbf{r}})$ therefore represents the time delay suffered by a particle scattered in the direction $\hat{\mathbf{r}}$, and is therefore referred to as the "delay time".

7.6. Cross-sections

We shall now derive the formula (1.2.9) connecting the differential cross-section with the scattering amplitude. The initial wave packet

$\bar{\varphi}(\mathbf{r}, t)$ is given by

$$\bar{\varphi}(\mathbf{r}, t) = (2\pi)^{-3/2} \int d\mathbf{k}\, C(\mathbf{k}) \exp(i\mathbf{k}\cdot\mathbf{r} - iE_k t/\hbar). \tag{7.6.1}$$

Since $\bar{\varphi}(t)$ is normalized to unity, and this holds for all times including $t = 0$, we must have

$$\int d\mathbf{r} |\bar{\varphi}(\mathbf{r}, 0)|^2 = 1. \tag{7.6.2}$$

Suppose that the packet has a cross-section A perpendicular to Oz; A will have the shape and area of the diaphragm through which the incident beam is released from the emitting apparatus. Equation (7.6.2) now becomes

$$\int_A\int dxdy \int_{-\infty}^{+\infty} dz |\bar{\varphi}(x, y, z, 0)|^2 = 1. \tag{7.6.3}$$

Now the integral with respect to z is a function of x and y which, when multiplied by $dxdy$, gives the probability of the particle P being in a cylinder of cross-section $dxdy$ and axis parallel to Oz. In a properly performed scattering experiment the incident beam is uniform, and so this probability must be independent of x and y. We can therefore put $x = y = 0$ so that (7.6.3) becomes

$$\int_A\int dxdy \int_{-\infty}^{+\infty} dz |\bar{\varphi}(0, 0, z, 0)|^2 = 1. \tag{7.6.4}$$

The integrand of the double integral in (7.6.4) is independent of x and y, and so if we denote the area of A by A we obtain

$$A \int_{-\infty}^{+\infty} dz |\bar{\varphi}(0, 0, z, 0)|^2 = 1. \tag{7.6.5}$$

If we substitute for $\bar{\varphi}(0, 0, z, 0)$ in (7.6.5) from (7.6.1) we have

$$1 = (2\pi)^{-3} A \int_{-\infty}^{+\infty} dz \int_{-\infty}^{+\infty} dk_x \int_{-\infty}^{+\infty} dk_y \int_{-\infty}^{+\infty} dk_z\, C^*(k_x, k_y, k_z) \exp(-ik_z z) \times$$

$$\times \int_{-\infty}^{+\infty} dk'_x \int_{-\infty}^{+\infty} dk'_y \int_{-\infty}^{+\infty} dk'_z\, C(k'_x, k'_y, k'_z) \exp(ik'_z z). \tag{7.6.6}$$

We can carry out the integration over z to obtain $2\pi\delta(k'_z - k_z)$, then integrate over k'_z, when we obtain

$$1 = A(2\pi)^{-2} \int_{-\infty}^{+\infty} dk_z \int_{-\infty}^{+\infty} dk_x \int_{-\infty}^{+\infty} dk_y\, C^*(k_x, k_y, k_z) \times$$

$$\times \int_{-\infty}^{+\infty} dk'_x \int_{-\infty}^{+\infty} dk'_y\, C(k'_x, k'_y, k_z)$$

$$= A(2\pi)^{-2} \int_{-\infty}^{+\infty} dk_z \left| \int_{-\infty}^{+\infty} dk_x \int_{-\infty}^{+\infty} dk_y\, C(k_x, k_y, k_z)\right|^2, \tag{7.6.7}$$

and so

$$\int_{-\infty}^{+\infty} dk_z \left| \int_{-\infty}^{+\infty} dk_x \int_{-\infty}^{+\infty} dk_y \, C(k_x, k_y, k_z) \right|^2 = (2\pi)^2 A^{-1}. \qquad (7.6.8)$$

Let us suppose that N particles are emitted per unit time. The incident flux is then $I = NA^{-1}$, and so $A^{-1} = IN^{-1}$. Hence (7.6.8) can be replaced by

$$\int_{-\infty}^{+\infty} dk_z \left| \int_{-\infty}^{+\infty} dk_x \int_{-\infty}^{+\infty} dk_y \, C(k_x, k_y, k_z) \right|^2 = 4\pi^2 \, IN^{-1}. \qquad (7.6.9)$$

The probability distribution for observing the state $\varphi_{\mathbf{l}} \exp(-iE_l t/\hbar)$ after the collision is $|A'(\mathbf{l})|^2$, where $A'(\mathbf{l}) \equiv \langle \varphi_{\mathbf{l}} \exp(-iE_l t/\hbar) | \bar{\varphi}'(t) \rangle$ is the probability amplitude. From (7.5.3) we see that $A'(\mathbf{l}) = D(\mathbf{l})$, and so from (7.5.5)

$$A'(\mathbf{l}) = C(\mathbf{l}) + \Delta(\mathbf{l}). \qquad (7.6.10)$$

Now we cannot measure the scattering amplitude for particles of momentum $\hbar\mathbf{l}$ where $C(\mathbf{l}) \neq 0$; for in doing so the detector would be swamped by the incident beam. Hence we need consider only the case $C(\mathbf{l}) = 0$, and if we put $\mathbf{l} = l\hat{\mathbf{r}}$, where $\hat{\mathbf{r}}$ is a unit vector in the direction of the scattered particle, (7.6.10) becomes

$$A'(l\hat{\mathbf{r}}) = \Delta(l\hat{\mathbf{r}}). \qquad (7.6.11)$$

In a well-performed experiment the collimation is sufficiently good to ensure that there is no variation in the scattering amplitude $f(l\hat{\mathbf{k}} \to l\hat{\mathbf{r}})$ as $\hat{\mathbf{k}}$ varies over the possible directions of motion in the incident beam, and so we can use (7.5.18) to replace (7.6.11) by

$$A'(l\hat{\mathbf{r}}) = (2\pi)^{-1} \, il \, f(l\hat{\mathbf{k}}_0 \to l\hat{\mathbf{r}}) \int d\hat{\mathbf{k}} \, C(l\hat{\mathbf{k}}). \qquad (7.6.12)$$

In order to obtain the differential cross-section we must first obtain the probability $P(\hat{\mathbf{r}})\,d\hat{\mathbf{r}}$ that the scattered particle emerges in the solid angle $d\hat{\mathbf{r}}$ in the direction $\hat{\mathbf{r}}$. The probability that the wave vector lies in the volume element $l^2 dl d\hat{\mathbf{r}}$ at the point $l\hat{\mathbf{r}}$ of $\mathbf{k}$-space is $|A'(l\hat{\mathbf{r}})|^2 \, l^2 dl d\hat{\mathbf{r}}$, and so the probability $P(\hat{\mathbf{r}})\,d\hat{\mathbf{r}}$ of the momentum being in the solid angle $(\hat{\mathbf{r}}, d\hat{\mathbf{r}})$ must be

$$P(\hat{\mathbf{r}})\,d\hat{\mathbf{r}} = \int_0^\infty l^2 \, dl d\hat{\mathbf{r}} |A'(l\hat{\mathbf{r}})|^2 = d\hat{\mathbf{r}} \int_0^\infty l^2 \, dl |A'(l\hat{\mathbf{r}})|^2. \qquad (7.6.13)$$

If we substitute for $A'(l\hat{\mathbf{r}})$ from (7.6.12) into (7.6.13) we have

$$P(\hat{\mathbf{r}}) = \int_0^\infty dl \, \frac{l^4}{4\pi^2} \, |f(l\hat{\mathbf{k}}_0 \to l\hat{\mathbf{r}})|^2 \, | \int d\hat{\mathbf{k}} \; C(l\hat{\mathbf{k}})|^2. \tag{7.6.14}$$

If the energy resolution is good $l^4|f|^2$ will be practically constant for values of l for which $C(l\hat{\mathbf{k}}) \neq 0$; we can therefore replace this quantity by its value when $l = k_0$ and take it outside the integral sign. Thus, (7.6.14) becomes

$$P(\hat{\mathbf{r}}) = (4\pi^2)^{-1} \, k_0^4 |f(\mathbf{k}_0 \to k_0\hat{\mathbf{r}})|^2 \int_0^\infty dl | \int d\hat{\mathbf{k}} \; C(l\hat{\mathbf{k}})|^2. \tag{7.6.15}$$

Now we are assuming that collimation is good, and so if we suppose $l\hat{\mathbf{k}}$ has components (k_x, k_y, k_z), $C(l\hat{\mathbf{k}})$ will vanish unless $k_z \simeq k_0$ and $k_x, k_y \ll k_0$. Thus Ox and Oy will be almost perpendicular to $l\hat{\mathbf{k}}$ if $C(l\hat{\mathbf{k}}) \neq 0$ (see Fig. 1.4), and so in the integral over $\hat{\mathbf{k}}$ in (7.6.15) we can replace the solid angle $d\hat{\mathbf{k}}$ by $dk_x \, dk_y/k_0^2$ to obtain

$$P(\hat{\mathbf{r}}) = (4\pi^2)^{-1} |f(\mathbf{k}_0 \to k_0\hat{\mathbf{r}})|^2 \int_0^\infty dl | \int_{-\infty}^{+\infty} dk_x \int_{-\infty}^{+\infty} dk_y \, C(k_x, k_y, k_z)|^2; \tag{7.6.16}$$

the limits are clearly $-\infty < k_x, k_y < +\infty$ since C vanishes except for $k_x, k_y \ll k_0$. We also have

$$l^2 = |l\hat{\mathbf{r}}|^2 = k_x{}^2 + k_y{}^2 + k_z{}^2, \tag{7.6.17}$$

and so for fixed l we can treat k_z as a function of k_x and k_y. If we then differentiate (7.6.17) implicitly with respect to k_x keeping k_y constant we have

$$0 = k_x + k_z \frac{\partial k_z}{\partial k_x} \tag{7.6.18}$$

and so

$$\left| \frac{\partial k_z}{\partial k_x} \right| = \left| \frac{k_x}{k_z} \right| \lesssim \frac{\Delta k_x}{k_0} \quad \text{for} \quad C(l\hat{\mathbf{k}}) \neq 0; \tag{7.6.19}$$

a similar expression holds when x is replaced by y. Since $\Delta k_x/k_0 \ll 1$ (7.6.19) shows that as k_x, k_y vary according to (7.6.17) k_z remains approximately constant and therefore equal to its value l when $k_x = k_y = 0$. Hence

$$\begin{aligned} &\int_0^\infty dl | \int_{-\infty}^{+\infty} dk_x \int_{-\infty}^{+\infty} dk_y \, C(k_x, k_y, k_z)|^2 \\ &= \int_0^\infty dl | \int_{-\infty}^{+\infty} dk_x \int_{-\infty}^{+\infty} dk_y \, C(k_x, k_y, l)|^2 \\ &= \int_0^\infty dk_z | \int_{-\infty}^{+\infty} dk_x \int_{-\infty}^{+\infty} dk_y \, C(k_x, k_y, k_z)|^2 \end{aligned} \tag{7.6.20}$$

since l is a dummy variable of integration. Since $C(\mathbf{k}) = 0$ if $k_z < 0$, we may replace the lower limit in the integration over k_z in (7.6.20) by $-\infty$ and then use (7.6.9) to obtain

$$\int_0^\infty dl \left| \int_{-\infty}^{+\infty} dk_x \int_{-\infty}^{+\infty} dk_y\, C(k_x, k_y, k_z) \right|^2 = 4\pi^2 IN^{-1}. \tag{7.6.21}$$

Substitution for the left-hand side of (7.6.21) in (7.6.16) yields

$$P(\hat{\mathbf{r}}) = |f(\mathbf{k}_0 \to k_0\hat{\mathbf{r}})|^2\, IN^{-1}. \tag{7.6.22}$$

Now N is the number of particles emitted per unit time, and so $NP(\hat{\mathbf{r}})\, d\hat{\mathbf{r}}$ is the number of particles emerging in the solid angle $d\hat{\mathbf{r}}$ per unit time. This divided by the incident flux I is $\sigma(\hat{\mathbf{r}})\, d\hat{\mathbf{r}}$ where $\sigma(\hat{\mathbf{r}})$ is the differential cross-section, and so

$$\sigma(\hat{\mathbf{r}}) = NI^{-1}\, P(\hat{\mathbf{r}}). \tag{7.6.23}$$

If we substitute for $P(\hat{\mathbf{r}})$ from (7.6.22) into (7.6.23) we obtain the result

$$\boxed{\sigma(\hat{\mathbf{r}}) = |f(\mathbf{k}_0 \to k_0\hat{\mathbf{r}})|^2} \tag{7.6.24}$$

which is identical with (1.2.9); but now the derivation has accurately simulated the experimental situation by following the behaviour of a wave packet travelling towards the target and being scattered by it.

7.7. The Pauli principle

In Sections 3.1 and 3.2 we showed that the problem of the scattering of two particles can be reduced to that of the scattering of a single particle by a centre of force, and in Section 3.3 we discussed the modifications which must be made to the treatment when the particles are identical. We pointed out there that the treatment given was not entirely satisfactory. Now that we have the time-dependent theory at our disposal, we are in a position to take up the discussion again and give a more satisfactory argument for the validity of the formula (3.3.13) for the differential cross-section.

Before we discuss the Pauli principle we shall investigate the properties of the "permutation operator" P and "symmetrizer" $\mathscr{S}$ for a system of two particles moving under a Hamiltonian $\mathsf{K} + V$ where K is the kinetic energy operator and V is a function only of the relative displacement $\mathbf{r}$. The permutation operator P is defined as the operator which interchanges the particles in the wave function. The centre of mass is unaffected by this, and the wave function ψ for relative motion is transformed by P into the wave function $\mathsf{P}\psi$ according to

$$\mathsf{P}\psi(\mathbf{r}) = \psi(-\mathbf{r}). \tag{7.7.1}$$

It follows that P satisfies

$$\mathsf{P}^2 = 1. \tag{7.7.2}$$

The symmetrizer $\mathscr{S}$ is defined by the expression

$$\mathscr{S} \equiv \tfrac{1}{2}(1+\mathsf{P}) \tag{7.7.3}$$

and from (7.7.2) and (7.7.3) we see that

$$\mathscr{S}^2 = \mathscr{S} \tag{7.7.4}$$

so that $\mathscr{S}$ has the properties of a projection operator (Volume 1, Appendix 4).

The symmetrizer is Hermitian, for if ψ, φ are wave functions

$$\begin{aligned}\langle\psi|\mathscr{S}|\varphi\rangle &= \int d\mathbf{r}\ \psi^*(\mathbf{r})\mathscr{S}\varphi(\mathbf{r}) \\ &= \tfrac{1}{2}\int d\mathbf{r}\ \psi^*(\mathbf{r})\ \varphi(\mathbf{r})+\tfrac{1}{2}\int d\mathbf{r}\ \psi^*(\mathbf{r})\ \varphi(-\mathbf{r}). \end{aligned}\tag{7.7.5}$$

If we make the change of variable $\mathbf{r} \to -\mathbf{r}$ in the second integral on the right-hand side of (7.7.5) we have

$$\begin{aligned}\langle\psi|\mathscr{S}|\varphi\rangle &= \tfrac{1}{2}\int d\mathbf{r}\ \psi^*(\mathbf{r})\ \varphi(\mathbf{r})+\tfrac{1}{2}\int d\mathbf{r}\ \psi^*(-\mathbf{r})\ \varphi(\mathbf{r}) \\ &= \int d\mathbf{r}\ \tfrac{1}{2}[\psi(\mathbf{r})+\mathsf{P}\psi(\mathbf{r})]^*\ \varphi(\mathbf{r});\end{aligned}$$

$$\therefore\ \langle\psi|\mathscr{S}|\varphi\rangle = \langle\mathscr{S}\psi|\varphi\rangle, \tag{7.7.6}$$

and (7.7.6) establishes the Hermiticity of $\mathscr{S}$.

The Pauli principle is a basic postulate of quantum mechanics. In the case of spinless particles it states that the wave functions found in nature are symmetric in the coordinates. This means that in the special case we are considering $\psi(-\mathbf{r}) = \psi(\mathbf{r})$, or equivalently, $\mathsf{P}\psi = \psi$. Now if φ is any wave function, symmetric or otherwise, it follows from (7.7.3) and (7.7.2) that

$$\mathsf{P}\mathscr{S}\varphi = \tfrac{1}{2}(\mathsf{P}+\mathsf{P}^2)\varphi = \tfrac{1}{2}(\mathsf{P}+1)\varphi = \mathscr{S}\varphi, \tag{7.7.7}$$

and so $\mathscr{S}\varphi$ is always a permissible wave function.

We shall now consider the effect of the Pauli principle on the theory of scattering of two spinless particles. Firstly we consider the symmetrization of the initial wave packet $\bar{\varphi}$. This is given by

$$\bar{\varphi}(t) = \int d\mathbf{k}\ C(\mathbf{k})\ \varphi_{\mathbf{k}} \exp\,(-iE_k t/\hbar), \tag{7.7.8}$$

and has the property that it vanishes unless $z < (v_0-\Delta v)t-m_z+n\pi/\Delta k_z$

[see (6.2.10)]. Since t is large and negative this means that $\bar{\varphi}(\mathbf{r}, t)$ vanishes for positive values of z, and so $P\bar{\varphi}(\mathbf{r}, t)$ must vanish for negative values of z. These observations enable us to calculate $\|\mathscr{S}\bar{\varphi}(t)\|$, for

$$\begin{aligned}\|\mathscr{S}\bar{\varphi}(t)\|^2 &= \langle\mathscr{S}\bar{\varphi}(t)|\mathscr{S}|\bar{\varphi}(t)\rangle \\ &= \langle\bar{\varphi}(t)|\mathscr{S}^2|\bar{\varphi}(t)\rangle \quad \text{(by (7.7.6))} \\ &= \langle\bar{\varphi}(t)|\mathscr{S}|\bar{\varphi}(t)\rangle \quad \text{(by (7.7.4))} \\ &= \tfrac{1}{2}\langle\bar{\varphi}(t)|\bar{\varphi}(t)\rangle + \tfrac{1}{2}\langle\bar{\varphi}(t)|\mathsf{P}|\bar{\varphi}(t)\rangle. \end{aligned} \tag{7.7.9}$$

Now $\langle\bar{\varphi}(t)|\bar{\varphi}(t)\rangle = \|\bar{\varphi}(t)\|^2 = 1$, while

$$\langle\bar{\varphi}(t)|\mathsf{P}|\bar{\varphi}(t)\rangle = \int d\mathbf{r}\; \bar{\varphi}^*(\mathbf{r}, t)\; \bar{\varphi}(-\mathbf{r}, t). \tag{7.7.10}$$

If z is positive $\bar{\varphi}^*(\mathbf{r}, t)$ vanishes, and if z is negative $\bar{\varphi}(-\mathbf{r}, t)$ vanishes. Hence the integral in (7.7.10) is zero so that $\bar{\varphi}(t)$ is orthogonal to $\mathsf{P}\bar{\varphi}(t)$, and (7.7.9) gives $\|\mathscr{S}\bar{\varphi}(t)\|^2 = \frac{1}{2}$. Thus the properly normalized and symmetrized initial wave packet is given by

$$\bar{\varphi}_{\mathscr{S}}(t) \equiv \sqrt{2}\cdot\mathscr{S}\bar{\varphi}(t). \tag{7.7.11}$$

Let us now consider the interpretation of $|\bar{\varphi}_{\mathscr{S}}(\mathbf{r}, t)|^2$. From (7.7.11) we have

$$2|\bar{\varphi}_{\mathscr{S}}(\mathbf{r}, t)|^2 = |2\mathscr{S}\bar{\varphi}(\mathbf{r}, t)|^2 \tag{7.7.12}$$

and so by (7.7.1) and (7.7.3)

$$2|\bar{\varphi}_{\mathscr{S}}(\mathbf{r}, t)|^2 = |\bar{\varphi}(\mathbf{r}, t) + \bar{\varphi}(-\mathbf{r}, t)|^2. \tag{7.7.13}$$

Since $\bar{\varphi}(\mathbf{r}, t) = 0$ if $z > 0$ and $\bar{\varphi}(-\mathbf{r}, t) = 0$ if $z < 0$ the product term vanishes and so (7.7.13) becomes

$$2|\bar{\varphi}_{\mathscr{S}}(\mathbf{r}, t)|^2 = |\bar{\varphi}(\mathbf{r}, t)|^2 + |\bar{\varphi}(-\mathbf{r}, t)|^2. \tag{7.7.14}$$

Initially the particles are distinguishable, being confined to different regions of space, and so $|\bar{\varphi}(\mathbf{r}, t)|^2\, d\mathbf{r}$ is the probability $P(\mathbf{r}, t)\, d\mathbf{r}$ that the relative displacement lies in a small volume $d\mathbf{r}$ at $\mathbf{r}$, while $|\bar{\varphi}(-\mathbf{r}, t)|^2$ $d\mathbf{r}$ is the probability that the relative displacement lies in a small volume $d\mathbf{r}$ at $-\mathbf{r}$. The sum of these is just the probability $P(\mathbf{r}, -\mathbf{r}, t)$ that the relative displacement lies in a volume $d\mathbf{r}$ either at $\mathbf{r}$ or at $-\mathbf{r}$; since $-\mathbf{r}$ corresponds to an interchange of the positions of the particles, $P(\mathbf{r}, -\mathbf{r}, t)\, d\mathbf{r}$ is just the probability that the displacement of one of the particles relative to the other lies in the volume $d\mathbf{r}$ at $\mathbf{r}$. We therefore have

$$P(\mathbf{r}, -\mathbf{r}, t) = 2|\bar{\varphi}_{\mathscr{S}}(\mathbf{r}, t)|^2. \tag{7.7.15}$$

The argument which led to (7.7.15) is only true initially when there is no overlap between $\bar{\varphi}$ and $\mathsf{P}\bar{\varphi}$ so that the particles are still distinguishable. We now take as an interpretive postulate the assumption that (7.7.15) holds at all times. Since $\bar{\varphi}_{\mathscr{S}}$ is normalized, the integral of $P(\mathbf{r}, -\mathbf{r}, t)$ over all $\mathbf{r}$ is equal to two. This is to be expected, however, for we count the probability $P(\mathbf{r}, -\mathbf{r}, t)$ twice; firstly as $P(\mathbf{r}, -\mathbf{r}, t)$, and then as $P(-\mathbf{r}, \mathbf{r}, t)$.

We must now seek the appropriate interpretive postulate for momentum. We first note that

$$\mathsf{P}\varphi_{\mathbf{k}}(\mathbf{r}) = \varphi_{\mathbf{k}}(-\mathbf{r}) = (2\pi)^{-3/2} \exp(-i\mathbf{k}\cdot\mathbf{r}) = \varphi_{-\mathbf{k}}(\mathbf{r}), \tag{7.7.16}$$

and so by (7.7.3)

$$\mathscr{S}\varphi_{\mathbf{k}} = \tfrac{1}{2}(\varphi_{\mathbf{k}} + \varphi_{-\mathbf{k}}). \tag{7.7.17}$$

Equation (7.7.8) now gives us

$$\begin{aligned}\bar{\varphi}_{\mathscr{S}}(t) &= \sqrt{2}\cdot\mathscr{S}\bar{\varphi}(t)\\ &= \frac{1}{\sqrt{2}}\int dk\, C(\mathbf{k})\,(\varphi_{\mathbf{k}} + \varphi_{-\mathbf{k}}) \exp(-iE_k t/\hbar),\end{aligned} \tag{7.7.18}$$

and so the probabilities of observing momentum $\hbar\mathbf{l}$ or $-\hbar\mathbf{l}$ are given by

$$|\langle\varphi_{\mathbf{l}} \exp(-iE_l\, t/\hbar)|\, \bar{\varphi}_{\mathscr{S}}(t)\rangle|^2 = \tfrac{1}{2}|C(\mathbf{l}) + C(-\mathbf{l})|^2, \tag{7.7.19}$$

$$|\langle\varphi_{-\mathbf{l}} \exp(-iE_l\, t/\hbar)|\, \bar{\varphi}_{\mathscr{S}}(t)\rangle|^2 = \tfrac{1}{2}|C(-\mathbf{l}) + C(\mathbf{l})|^2. \tag{7.7.20}$$

The probability $P(\mathbf{l}, -\mathbf{l}, t)$ of observing the particles having momentum $\hbar\mathbf{l}$ or momentum $-\hbar\mathbf{l}$ (when their relative velocity is reversed) is the sum of these, and is therefore $|C(\mathbf{l}) + C(-\mathbf{l})|^2$. We therefore get

$$P(\mathbf{l}, -\mathbf{l}, t) = 2|\langle\varphi_{\mathbf{l}} \exp(-iE_l\, t/\hbar)|\, \bar{\varphi}_{\mathscr{S}}(t)\rangle|^2; \tag{7.7.21}$$

this is true when the particles are distinguishable, as at times before the collision, and we postulate that it holds at all times. Again the integral of the probability over all $\mathbf{l}$-space is two, and for the same reason; each direction is counted twice.

Let us now consider the evolution in time of the state $\bar{\varphi}_{\mathscr{S}}(t)$. It evolves in time into the state $\psi_{\mathscr{S}}(t)$ where

$$\begin{aligned}\bar{\psi}_{\mathscr{S}}(t) &= \exp[-i\mathsf{H}(t-t_0)/\hbar]\, \bar{\varphi}_{\mathscr{S}}(t_0)\\ &= \exp[-i\mathsf{H}(t-t_0)/\hbar]\sqrt{2}\cdot\mathscr{S}\bar{\varphi}(t_0).\end{aligned} \tag{7.7.22}$$

Now the Hamiltonian H which governs the motion of two identical particles must be invariant under the interchange of their coordinates, for otherwise there would be a means of distinguishing them. The Hamiltonian must therefore commute with the symmetrizer $\mathscr{S}$, and so the evolution operator must also commute with $\mathscr{S}$. This means that (7.7.22) can be written

$$\bar{\psi}_{\mathscr{S}}(t) = \sqrt{2}\cdot\mathscr{S} \exp\left[-i\mathsf{H}(t-t_0)/\hbar\right] \bar{\varphi}(t_0). \tag{7.7.23}$$

Equation (7.7.23) is more convenient to consider than (7.7.22), for we need not symmetrize the initial state, but instead we let it evolve and then carry out the symmetrization after the collision. Since we have already dealt with the problem of the evolution of $\bar{\varphi}(t_0)$, we need only use the results already obtained. In addition, we avoid any difficulties with interpretation of flux. Initially the particles are distinguishable, and so it is logical to consider the initial state as $\bar{\varphi}(t)$. We must symmetrize *after* the collision, when the particles are no longer distinguishable, for there is no means of determining which is the scattered particle and which is the recoil particle.

Since $\bar{\varphi}(t_0)$ evolves into $\bar{\psi}(t)$, (7.7.23) may be replaced by

$$\bar{\psi}_{\mathscr{S}}(t) = \sqrt{2}.\mathscr{S}\bar{\psi}(t). \tag{7.7.24}$$

After the collision the state $\bar{\varphi}(t_0)$ has evolved into the state $\bar{\varphi}'(t_1)$, where t_1 is a sufficiently large and positive time, and so from (7.7.24) the state $\bar{\psi}_{\mathscr{S}}(t)$ has evolved into the state $\bar{\varphi}'_{\mathscr{S}}(t_1)$ where

$$\bar{\varphi}'_{\mathscr{S}}(t_1) = \sqrt{2}\cdot\mathscr{S}\bar{\varphi}'(t_1). \tag{7.7.25}$$

The probability $P(\mathbf{l}, -\mathbf{l})$ of the final relative momentum being either $\hbar\mathbf{l}$ or $-\hbar\mathbf{l}$ is therefore given by

$$P(\mathbf{l}, -\mathbf{l}) = |A'(\mathbf{l}, -\mathbf{l})|^2 \tag{7.7.26}$$

where, since (7.7.21) is postulated to hold for any normalized symmetrized wave function $\bar{\varphi}_{\mathscr{S}}(t)$, and therefore in particular when $\bar{\varphi}_{\mathscr{S}}(t)$ is replaced by $\bar{\varphi}'_{\mathscr{S}}(t)$,

$$A'(\mathbf{l}, -\mathbf{l}) = \sqrt{2}\cdot\langle\varphi_{\mathbf{l}} \exp(-iE_l t/\hbar)|\ \bar{\varphi}'_{\mathscr{S}}(t_1)\rangle; \tag{7.7.27}$$

hence by (7.7.25),

$$A'(\mathbf{l}, -\mathbf{l}) = 2\langle\varphi_{\mathbf{l}} \exp(-iE_l t_1/\hbar)|\ \mathscr{S}\bar{\varphi}'(t_1)\rangle. \tag{7.7.28}$$

If we use (7.7.6) and (7.7.17) we see that (7.7.28) becomes

$$\begin{aligned} A'(\mathbf{l}, -\mathbf{l}) &= \langle\varphi_{\mathbf{l}} \exp(-iE_l t/\hbar) + \varphi_{-\mathbf{l}} \exp(-iE_l t(\hbar)|\ \bar{\varphi}'(t_1)\rangle \\ &= A'(\mathbf{l}) + A'(-\mathbf{l}) \end{aligned} \tag{7.7.29}$$

where $A'(\mathbf{l})$ is the amplitude considered in Section 7.6. Hence the probability amplitude $A'(l\hat{\mathbf{r}}, -l\hat{\mathbf{r}})$ for *a* particle emerging in the final state in the direction $\hat{\mathbf{r}}$ relative to the other one is

$$A'(l\hat{\mathbf{r}},\ -l\hat{\mathbf{r}}) = A'(l\hat{\mathbf{r}})+A'(-l\hat{\mathbf{r}}), \tag{7.7.30}$$

and so from (7.6.12) this is

$$A'(l\hat{\mathbf{r}},\ -l\hat{\mathbf{r}}) = (2\pi)^{-1}\, il[f(l\hat{\mathbf{k}}_0 \rightarrow l\hat{\mathbf{r}})+f(l\hat{\mathbf{k}}_0 \rightarrow -l\hat{\mathbf{r}})] \int d\mathbf{k}\ C(l\hat{\mathbf{k}}). \tag{7.7.31}$$

The argument is now identical with that of Section 7.6 with the amplitude $f(l\hat{\mathbf{k}}_0 \rightarrow l\hat{\mathbf{r}})+f(l\hat{\mathbf{k}}_0 \rightarrow -l\hat{\mathbf{r}})$ replacing the amplitude $f(l\hat{\mathbf{k}}_0 \rightarrow l\hat{\mathbf{r}})$, and so (7.6.24) is replaced by (since $l = k_0$)

$$\boxed{\sigma(\hat{\mathbf{r}}) = |f(\mathbf{k}_0 \rightarrow k_0\hat{\mathbf{r}})+f(\mathbf{k}_0 \rightarrow -k_0\hat{\mathbf{r}})|^2.} \tag{7.7.32}$$

The result (7.7.32) is identical with (3.3.13), which is what we set out to prove.

APPENDIX A

The optical theorem

At those points where the scattering wave function has reached its asymptotic form (1.2.11), the radial component J_r of the current vector **J** is given by [cf. (1.2.6)]

$$
\begin{aligned}
J_r = \frac{\hbar}{2i\mu}\Bigg\{&\left[\exp(-ikr\cos\theta)+f^*(\theta,\varphi)\frac{\exp(-ikr)}{r}\right]\times\\
&\times\frac{\partial}{\partial r}\left[\exp(ikr\cos\theta)+f(\theta,\varphi)\frac{\exp(ikr)}{r}\right]\\
&-\left[\exp(ikr\cos\theta)+f(\theta,\varphi)\frac{\exp(ikr)}{r}\right]\times\\
&\times\frac{\partial}{\partial r}\left[\exp(-ikr\cos\theta)+f^*(\theta,\varphi)\frac{\exp(-ikr)}{r}\right]\Bigg\}\\
= \frac{\hbar}{2i\mu}\Bigg\{&\left[\exp(-ikr\cos\theta)+f^*(\theta,\varphi)\frac{\exp(-ikr)}{r}\right]\times\\
&\times\left[ik\cos\theta\exp(ikr\cos\theta)+ikf(\theta,\varphi)\frac{\exp(ikr)}{r}-f(\theta,\varphi)\frac{\exp(ikr)}{r^2}\right]\\
&-\left[\exp(ikr\cos\theta)+f(\theta,\varphi)\frac{\exp(ikr)}{r}\right]\times\\
&\times\Bigg[-ik\cos\theta\exp(-ikr\cos\theta)-ikf^*(\theta,\varphi)\frac{\exp(-ikr)}{r}\\
&-f^*(\theta,\varphi)\frac{\exp(-ikr)}{r^2}\Bigg]\Bigg\}\\
= \frac{\hbar k}{\mu}\Bigg\{&\cos\theta+(\cos\theta+1)\,\mathrm{Re}\left[\frac{f(\theta,\varphi)\exp ikr(1-\cos\theta)}{r}\right]\\
&+\frac{|f(\theta,\varphi)|^2}{r^2}\Bigg\}-\frac{\hbar}{\mu}\,\mathrm{Im}\left[f(\theta,\varphi)\frac{\exp ikr(1-\cos\theta)}{r^2}\right]+0\left(\frac{1}{r^3}\right).
\end{aligned}
\tag{A1}
$$

Since particles are conserved, if S is a large sphere of radius r and centre the origin

$$\int_S J_r \, dS = r^2 \int_0^{2\pi} d\varphi \int_0^{\pi} \sin\theta \, d\theta \, J_r = 0. \tag{A2}$$

If we substitute (A1) into (A2), we note that the θ-integration of the second term gives, if $t = \cos\theta$,

$$\frac{1}{r} \operatorname{Re} \int_{-1}^{+1} (t+1) f(\theta, \varphi) \exp\{ikr(1-t)\} \, dt$$

$$= \operatorname{Re}\left[-\frac{1}{ikr^2}(t+1) f(\theta, \varphi) \exp\{ikr(1-t)\}\right]_{t=-1}^{t=+1}$$

plus a term which can be shown to be order of r^{-3} on repeated integration by parts; hence as $r \to \infty$ this behaves as

$$-\frac{2}{kr^2} \operatorname{Im} f(0, \varphi) = -\frac{2}{kr^2} \operatorname{Im} f(0).$$

Integrating over φ, this $= -4\pi/kr^2 \operatorname{Im} f(0)$.

The fourth term in (A2), on integration by parts over θ, is seen to be of order r^{-3} and the first term vanishes on integration over θ. Hence

$$\frac{\hbar k}{\mu}\left\{-\frac{4\pi}{k} \operatorname{Im} f(0) + \int_0^{2\pi} d\varphi \int_0^{\pi} \sin\theta \, d\theta |f(\theta, \varphi)|^2\right\} = 0,$$

which is the Optical Theorem (1.4.30).

APPENDIX B

To evaluate

$$V_{00}(\mathbf{r}) \equiv \int |\chi_0(\boldsymbol{\rho})|^2 \; V(\mathbf{r}, \boldsymbol{\rho}) \, d\boldsymbol{\rho}. \qquad \text{(B1)}$$

If we substitute for V from (4.2.3) and remember that χ_0 is normalized we find that

$$V_{00}(\mathbf{r}) = -\frac{e^2}{\kappa_0 r} + e^2 \int |\chi_0(\boldsymbol{\rho})|^2 \frac{1}{\kappa_0 |\mathbf{r}-\boldsymbol{\rho}|} \, d\boldsymbol{\rho}. \qquad \text{(B2)}$$

If we use (4.3.34) we see that (using spherical polar coordinates with $\mathbf{r}$ as polar axis)

$$\int d\boldsymbol{\rho} |\chi_0(\boldsymbol{\rho})|^2 \frac{1}{|\mathbf{r}-\boldsymbol{\rho}|}$$

$$= (\pi a_0{}^3)^{-1} \int_0^\infty \rho^2 \, d\rho \int_{-1}^{+1} dt \int_0^{2\pi} d\varphi \exp(-2\rho/a_0) \, (r^2+\rho^2-2r\rho t)^{-\frac{1}{2}}$$

$$= -2a_0^{-3} \, r^{-1} \int_0^\infty \rho[(r^2+\rho^2-2r\rho t)^{\frac{1}{2}}]_{t=-1}^{t=+1} \exp(-2\rho/a_0) \, d\rho$$

$$= -2a_0^{-3} \, r^{-1} \{ \int_0^r \rho[(r-\rho)-(r+\rho)] \exp(-2\rho/a_0) \, d\rho$$

$$+ \int_r^\infty \rho[(\rho-r)-(\rho+r)] \exp(-2\rho/a_0) \, d\rho$$

$$= -a_0^{-1} \exp(-2r/a_0) + r^{-1} - r^{-1} \exp(-2r/a_0). \qquad \text{(B3)}$$

Substitution of (B3) into (B2) gives

$$V_{00}(\mathbf{r}) = -e^2 \kappa_0^1 (a_0^{-1} + r^{-1}) \exp(-2r/a_0) \qquad \text{(B4)}$$

which is (4.2.37).

APPENDIX C

Contour integration

A complex-valued function $f(z)$ of the complex variable z is analytic at a point z if the limit

$$\lim_{h \to 0} \frac{f(z+h)-f(z)}{h}$$

exists and is unique. The following theorem, known as the theorem of residues, is proved in books on complex variable:

"Let $f(z)$ be analytic and single valued inside and on a simple closed contour C, except at a finite number of singularities at $z_1, z_2, \ldots, z_n$, and let the residues of $f(z)$ at these points be $R_1, R_2, \ldots, R_n$. Then

$$\int_C f(z)\, dz = 2\pi i \sum_{m=1}^{n} R_m."$$

In the above theorem the residue R_m at the point z_m is defined by

$$R_m = \lim_{z \to z_m} (z-z_m)\, f(z),$$

and we assume that this exists for each z_m.

Jordan's lemma may be stated as follows:

"Let $f(z)$ be analytic and let $|f(z)| \to 0$ uniformly as $|z| \to \infty$, when $|z| > c > 0$, and Im $(z) \geqq 0$. Further, let m be a positive constant and let Γ be a semicircle or radius $R > c$ above the real axis with centre as origin. Then

$$\lim_{R \to \infty} \int_{\Gamma} \exp\,(miz)\, f(z)\, dz = 0."$$

(By $|f(z)| \to 0$ uniformly we mean that $|f(z)| \leqq U_R$ on Γ, where $U_R \to 0$ as $R \to \infty$.)

In the example of Section 5.1, $f(z)$ is given by

$$f(z) = (k^2 - z^2 + i\eta)^{-1} z,$$

$$\therefore\ |f(z)| = |k^2 - z^2 + i\eta|^{-1} |z|;$$

on the semicircle $|z| = R$, and the minimum value of $|k^2 - z^2 + i\eta|$ occurs when z^2 is parallel to $k^2+i\eta$, when it is equal to $R^2-|k^2+i\eta|$. Thus

$$|f(z)| \leqq \frac{R}{R^2-|k^2+i\eta|},$$

and this clearly tends to zero as $R \to \infty$. It therefore follows that the theorem of residues may be applied in this case, which justifies the procedure of Section 5.1.

APPENDIX D

Evaluation of the operator integrals

$$\mathsf{I}^+ \equiv \int_0^{-\infty} d\tau \exp\,[i(\mathsf{H}-E_k-i\varepsilon)\,\tau/\hbar],$$

$$\mathsf{I}^- \equiv \int_0^{+\infty} d\tau \exp\,[i(\mathsf{H}-E_k+i\varepsilon)\,\tau/\hbar] \quad (\varepsilon > 0).$$

If we introduce a complete set of orthonormal eigenfunctions ψ_ν of H, we see that if ψ is any wave function,

$$\mathsf{I}^+\psi = \int_0^{-\infty} d\tau \sum_\nu \exp\,[i(E_\nu-E_k-i\varepsilon)\,\tau/\hbar]\,\psi_\nu\langle\psi_\nu|\psi\rangle.$$

Due to the presence of the term $\varepsilon\tau/\hbar$ in the exponential the integrand tends to zero exponentially as $\tau \rightarrow -\infty$. We can therefore interchange the order of integration to obtain

$$\mathsf{I}^+\psi = \sum_\nu \frac{\hbar}{i}\,(E_k-E_\nu+i\varepsilon)^{-1}\,\psi_\nu\langle\psi_\nu|\psi\rangle$$

and from the usual rule for expanding a function of H we see that

$$\mathsf{I}^+ = \frac{\hbar}{i}\,(E_k-\mathsf{H}+i\varepsilon)^{-1}.$$

In a similar way we may show that

$$\mathsf{I}^- = \frac{\hbar}{i}\,(E_k-\mathsf{H}-i\varepsilon)^{-1}.$$

INDEX